Springer Theses

Recognizing Outstanding Ph.D. Research

Aims and Scope

The series "Springer Theses" brings together a selection of the very best Ph.D. theses from around the world and across the physical sciences. Nominated and endorsed by two recognized specialists, each published volume has been selected for its scientific excellence and the high impact of its contents for the pertinent field of research. For greater accessibility to non-specialists, the published versions include an extended introduction, as well as a foreword by the student's supervisor explaining the special relevance of the work for the field. As a whole, the series will provide a valuable resource both for newcomers to the research fields described, and for other scientists seeking detailed background information on special questions. Finally, it provides an accredited documentation of the valuable contributions made by today's younger generation of scientists.

Theses may be nominated for publication in this series by heads of department at internationally leading universities or institutes and should fulfill all of the following criteria

- They must be written in good English.
- The topic should fall within the confines of Chemistry, Physics, Earth Sciences, Engineering and related interdisciplinary fields such as Materials, Nanoscience, Chemical Engineering, Complex Systems and Biophysics.
- The work reported in the thesis must represent a significant scientific advance.
- If the thesis includes previously published material, permission to reproduce this must be gained from the respective copyright holder (a maximum 30% of the thesis should be a verbatim reproduction from the author's previous publications).
- They must have been examined and passed during the 12 months prior to nomination.
- Each thesis should include a foreword by the supervisor outlining the significance of its content.
- The theses should have a clearly defined structure including an introduction accessible to new PhD students and scientists not expert in the relevant field.

Indexed by zbMATH.

More information about this series at http://www.springer.com/series/8790

Giuseppe Ferrara

Flax-TRM Composite Systems for Strengthening of Masonry

From Material Identification to Structural Behavior

Doctoral Thesis accepted by
University of Salerno, Salerno, Italy & University of Lyon, Lyon, France

UNIVERSITÉ UNIVERSITÀ
FRANCO ITALO
ITALIENNE FRANCESE

Springer

Author
Dr. Giuseppe Ferrara
Department of Civil Engineering
University of Salerno
Fisciano, Italy

Laboratory of Composite Materials
for Constructions
University Claude Bernard Lyon 1
Villeurbanne, France

Supervisors
Prof. Enzo Martinelli
Department of Civil Engineering
University of Salerno
Fisciano, Italy

Prof. Aron Gabor
Laboratory of Composite Materials
for Constructions
University Claude Bernard Lyon 1
Villeurbanne, France

Prof. Carmelo Caggegi
Laboratory of Composite Materials
for Constructions
University Claude Bernard Lyon 1
Villeurbanne, France

ISSN 2190-5053 ISSN 2190-5061 (electronic)
Springer Theses
ISBN 978-3-030-70272-4 ISBN 978-3-030-70273-1 (eBook)
https://doi.org/10.1007/978-3-030-70273-1

This Springer imprint is published by the registered company Springer Nature Switzerland AG
The registered company address is: Gewerbestrasse 11, 6330 Cham, Switzerland

Research Funding

Several institutions financially contributed to the research study:

- the Italian Ministry for Education, University and Research (MIUR), as part of the program "Dottorati Innovativi a caratterizzazione industriale" funded by the European Union (Structural Funding ERDF-ESF for "Research and Innovation" 2014–2020), awarded the Ph.D. scholarship;
- the French Government provided a scholarship to cover the mobility expenses concerning the stay at the laboratory LMC2 (UCBL1, France);
- the Ph.D. school MEGA (UCBL1, France) contributed to the mobility expenses for participation to international conferences;
- the laboratory LMC2 (UCBL1, France) contributed to the mobility expenses for participation to international conferences;
- the European Union's Horizon 2020 Research and Innovation Program under Grant Agreement N° 645704 (H2020-MSCA-RISE-2014), by means of the "SUPERCONCRETE Project" (www.superconcrete-h2020.unisa.it), covered the mobility expenses concerning the stay at the laboratory NUMATS (UFRJ, Brazil);
- the company INNOVATIONS s.r.l. provided the materials tested in the experimental research activity;
- the "Université Franco-Italienne" contributed to the dissemination expenses of the research results by awarding the "UFI/UIF 2020" prize.

The author gratefully acknowledges the mentioned institutions for their financial support, essential for the fulfilment of the Thesis.

Supervisor's Foreword

It is a great pleasure to introduce Dr. Giuseppe Ferrara's Ph.D. dissertation, accepted for publication as part of the Springer Theses book series.

Dr. Ferrara joined my research group at the Department of Civil Engineering (DICiv) of the University of Salerno (UniSa) in 2013, as he carried out his Bachelor thesis, and, two years later, he continued his collaboration during his Master Thesis project, realised in partnership with Prof. Carmelo Caggegi of the University Claude Bernard Lyon 1 (France).

In 2016, Dr. Ferrara enrolled at the Ph.D. programme in "*Risk and sustainability in civil engineering, environmental and construction*" (Cycle XXXII) at the Department of Civil Engineering of the University of Salerno. The Ph.D. scholarship was funded by the Italian Ministry for Education, University and Research (MIUR) as part of the programme "Dottorati Innovativi a caratterizzazione industriale" financed by the European Union (Structural Funding ERDF-ESF for Research and Innovation 2014–2020). At the same time, he took the opportunity to enroll also at the Ph.D. school in "*Mechanic, Energetic, Civil Engineering and Acoustic*" (ED 162 MEGA), thanks to a bi-lateral agreement for a co-supervised Ph.D. Thesis between the University of Salerno and the University Claude Bernard Lyon 1 (France).

Therefore, Giuseppe Ferrara developed his Ph.D. research at both the Structural Engineering Testing Hall (Str.Eng.T.H.) of the Italian Institution and the Laboratory of Composite Materials for Construction (LMC2) of the French one. Moreover, Dr. Ferrara was also a visiting Ph.D. student at the NUMATS laboratory of the Federal University of Rio de Janeiro (Brazil), as part of the EU funded project "SUPERCONCRETE" (H2020-MSCA-RISE-2014 n. 645704).

Dr. Ferrara defended his theses on 6 April 2020, and he obtained the Ph.D. degree at both University of Salerno and University of Lyon. The committee awarded Dr. Ferrara the highest honour and, due to the clear international nature of the work, he also earned the "*Doctor Europaeus*" title.

This book is largely based on the contents of the Giuseppe Ferrara's Ph.D. Thesis. It investigates the behaviour of new type of composites, generally referred to as Textile Reinforced Mortars (TRMs), made with plant (namely, Flax) textiles. Specifically, the work proposes experimental results and theoretical analyses about both durability performance and mechanical behaviour of Flax-TRM systems, which demonstrate the potential of this innovative and sustainable composite system for seismic strengthening of existing masonry buildings.

Cross-scale investigations are reported in the Ph.D. Thesis, as the research moves from the study of the single flax fibres, covers the relevant physical and mechanical aspects of textiles and mortars and ends up presenting the results of full-scale diagonal shear tests on masonry walls externally strengthened by Flax-TRM systems. The experimental tests are carried out according to internationally recognised standards and executed by using state-of-the-art testing and measurement techniques.

Finally, the work proposes a comprehensive report on the physical properties and the structural potential of Flax-TRM systems comprising plant fibres, as well as it figures out possible future developments for both the composite system and the strengthening technique. I am convinced that Giuseppe will keep contributing with his passion and competence for the further advances of this research subject.

Fisciano, Italy
November 2020

Prof. Enzo Martinelli

Parts of this thesis have been published in the following articles:

Ferrara G., Caggegi C., Martinelli E., and Gabor A. (2020). Shear capacity of masonry walls externally strengthened using Flax-TRM composite systems: experimental tests and comparative assessment. *Construction and Building Materials*, 261, 120490.

Ferrara G., Coppola B, Di Maio L., Incarnato L., and Martinelli E. (2019). Tensile strength of flax fabrics to be used as reinforcement in cement-based composites: experimental tests under different environmental exposures, *Composites Part B: Engineering*, 168, 511–523.

Ferrara G., Pepe M., Martinelli E., and Toledo Filho R. D. (2019). Influence of an Impregnation Treatment on the Morphology and Mechanical Behaviour of Flax Yarns Embedded in Hydraulic Lime Mortar, *Fibers*, 7(4), 30.

Ferrara G., Caggegi C., Gabor A., and Martinelli E. (2019). Mechanical Behaviour of Masonry Panels Strengthened by Flax TRM Systems. *Key Engineering Materials*, 817, 427–434.

Ferrara G., and Martinelli E. (2018). Tensile behaviour of textile reinforced mortar composite systems with flax fibres. *Proceedings of the 12th fib International Ph.D. Symposium in Civil Engineering*, 863–869.

Acknowledgments

I would like to express my deepest gratitude and appreciation to the people without whom the completion of the Ph.D. thesis would not have been possible.

My sincere thanks to

Prof. Enzo Martinelli, my first mentor and the person who introduced me to the academic research world, for guiding me during the research, for the wise advices, for the constant encouragement, and for always believing in me;

Profs. Aron Gabor and Carmelo Caggegi, my French supervisors, for their fundamental guide and support during my research activity at University C.B. Lyon1, and for making me comfortable within the institution from the very first moment;

Mr. Rossano Petruzziello, the head of the company Engineering s.r.l. (Avellino, Italy), for giving me the possibility to spend time in his company;

Prof. Emmanuel Ferrier, the director of the laboratory LMC2 (University C.B. Lyon1, France), for gently welcoming me to the laboratory and for giving me the great opportunity to join it;

Prof. Luciano di Maio and Dr. Bartolomeo Coppola, who work at the Department of Industrial Engineering of University od Salerno, for their precious collaboration in fulfilling part of the experimental activity;

Prof. Romildo Dias Toledo Filho, director of the laboratory NUMATS (Federal University of Rio de Janeiro, Brazil), for kindly welcoming me to the laboratory, and giving me the great chance to cooperate with it;

Dr. Marco Pepe, researcher at University of Salerno, for the valuable support, and for the collaboration in part of the research.

I am also grateful to the entire staffs of Ph.D. students, technicians, secretaries, researchers and professors of both the laboratories LMC2 and NUMATS, for their hospitality and for treating me like a family.

Contents

About the Author

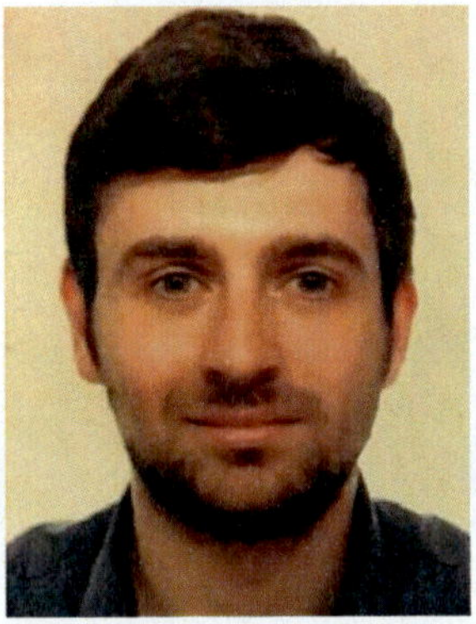

Giuseppe Ferrara was born in Italy, in 1991. He received his B.Sc. and M.Sc. degrees, with 110/110 cum laude, in Civil Engineering from the University of Salerno, Italy, in 2013 and 2016, respectively. In the period June–August 2016, he was a trainee at the Federal Institute for Materials Research and Testing (BAM) in Berlin (Germany), dealing with the study of Ultra-High performance fibre-reinforced concrete. In November 2016, he started a bi-nationally supervised Ph.D. course between the University of Salerno (Italy), at the Department of Civil Engineering—doctoral school in Risk and Sustainability in Civil, Architectural and Environmental Engineering (Cycle XXXII)—and the University Claude Bernard Lyon 1 (France), at the Laboratory of Composite Materials for Constructions (LMC^2)—doctoral school in Mechanic, Energetic, Civil Engineering and Acoustic (ED 162, MEGA). In 2018, he was a visiting Ph.D. student at the Federal University of Rio de Janeiro (Brazil) in the framework of the "*SUPERCONCRETE*" European project (H2020-MSCA-RISE-2014). On 6 April 2020, he defended his thesis obtaining the Ph.D. degree at both University of Salerno and University of Lyon. He was awarded the highest honour, and he also earned the title of "*Doctor Europaeus*". He was awarded the "*Prix UFI 2020*" from the "*Université Franco-Italienne*" for the thesis work. From July 2020, he is a post-doctoral research fellow at Laboratory of Composite Materials for Constructions (University CB Lyon1). In November 2020, he was awarded a grant for a Short Terms Scientific Mission carried out at University of Salerno in the framework of the COST Action "*European Network to connect research and innovation efforts on Advanced Smart Textiles (CONTEXT), CA17107*". His research deals with composites materials for strengthening of masonry structures, the use of plant fibres in composites, and timber-concrete composite structures. He published some of the achieved results on International Journals and presented his work in several International Conferences.

Chapter 1
Introduction

1.1 Motivation

The Mediterranean basin is characterised by a significant seismic hazard, due to geographic and morphological aspects. The many earthquakes events happened over centuries, have affected human activities with consequences in terms of safety, economy, and sociology. Nowadays, although a deep and established knowledge of the problem, and available technologies to conceive civil structures able to cope successfully with earthquake actions, seismic events still represent an actual threat for human safety. This concern is mainly due to the presence on the territory of several structures the construction of which often dates way back before the establishment of designing principles and methods specifically conceived to address such seismic issues [1].

The majority of the historical city centres are characterised by old masonry buildings, conceived to face gravitational loads, but very often lacking in capacity to resist to the horizontal forces caused by seismic events. The seismic hazard, combined with a massive presence of vulnerable structures, confers to the area a high seismic risk, against which it is necessary to provide urgent solutions to guarantee at least the preservation of human life [3, 4].

The increased awareness toward the preservation of our environment, represents nowadays an aspect of fundamental importance that affects more and more our society at every level. Ecological solutions are strongly recommended, and in some cases imposed, by ruling governments, in every production process of our community, with the aim to achieve, in a more or less far future, a circular economy system based on the use of renewable sources, recycling processes, and the reduction, as much as possible, and at any level, of the environmental impact.

In this framework, the field of the building materials is fully involved in such epoch-making phase of transition. Manufacturing companies, practitioners and scientific research groups are constantly looking for innovative solutions that may meet both mechanical and ecological requirements, giving rise to a new class of systems referred to as green building materials [7].

G. Ferrara, *Flax-TRM Composite Systems for Strengthening of Masonry*, Springer Theses,
https://doi.org/10.1007/978-3-030-70273-1_1

Inorganic matrix composites, consisting in high strength fibres embedded in cementitious or lime mortars (referred to as Textile Reinforced Mortar, TRM), represents an innovative and efficient strengthening technique to enhance the capacity of structural elements, very suitable in the case of masonry buildings [5]. With the aim of combining the aspects aforementioned, the research focuses on the study of a more sustainable TRM composite, and on the investigation of its efficiency as reinforcement of masonry structural elements.

The sustainability of the system lies in the use of plant fibres, instead of high strength synthetic ones. In fact, several plant fibres, such as hemp, flax, sisal, jute, coconut, curaua, etc...., already gained attention as reinforcement in composites in several fields (aeronautic, naval, automotive, etc....) thanks to several good properties, such as: availability, satisfactory mechanical characteristics, affordability, high resistance over density ratio, sustainability [6]. In recent years few studies addressed the problem of the use of plant-based composites also in the building material field. The use of plant fibre-based textiles in TRM composites showed up as a very promising reinforcement system with a great potential, but still characterised by several issues that prevent their diffusion on the market in the same manner of most traditional synthetic-based composites [2]. As a consequence, it is necessary to intensify the research in the field in order to deeply define the mechanical behaviour of such innovative composite systems, and to improve their efficiency according to the performance required in structural applications.

In order to comply with this need, the present study strives to provide a comprehensive investigation of an innovative sustainable plant based TRM system, i.e. Flax TRM, aiming at defining its mechanical behaviour, and at identifying its efficiency and limitations in strengthening of masonry structural elements.

In order to have the elements to deeply understand the behaviour of the system, the analysis starts from the investigation on a small scale of the composite constituent materials, to move until an experimental study on the structural element scale of investigation.

In addition, the issue of the durability of the composite was addressed as well, being the latter one of the main aspects that feed scepticism about the use of plant fibre-based composites [8].

Finally, the study aims at showing that there is still room for improvement with respect to the mechanical behaviour of plant based-TRMs, by proposing the application of some improving techniques conceived on the basis of the outcomes deriving from the reach multi-scale analysis proposed.

1.2 Contents and Structure of the Thesis

The thesis consists of the analysis of results deriving from experimental investigations carried out at different scales of study. From a small scale of investigation aimed at analysing the microstructure of flax fibres, passing through the assessment of the mechanical response of TRM composites, up to the study of masonry structural elements externally strengthened by Flax TRMs, the thesis provides a comprehensive identification of the behaviour of plant based composites.

Each chapter deals with a different scale of investigation. A critical analysis comparing the results at different scales of investigation is proposed as well. Finally, some modifications of the TRM components, aimed at improving the performance of the reinforcement, are experienced as conclusion of the work.

The structure of the thesis, with the main contents and outcomes of each section, are reported as follow:

- In this chapter presents the context of the research. It comprises the main contents and the aim of the thesis;
- Chapter 2 presents a review of the literature state of the art, concerning the topic of interest, is presented. Specifically, the main research outcomes available in the literature, are presented with respect to the of TRM composites as reinforcement systems in civil engineering, and with respect to the use of plant fibres in TRMs. The aims of the thesis are reported as well;
- Chapter 3 reports the outcomes of the investigation of the flax textile adopted in the research. It aims, by providing the main physical and mechanical properties of the textile, and its durability performance, at defining whether or not the fabric is suitable as reinforcement in mortar-based composites;
- Chapter 4 proposes the characterisation of the system on the composite scale of analysis. According to standardised procedures, and with respect to different reinforcement configurations proposed within the research, the composite mechanical behaviour was assessed by means of Flax TRM tensile tests, and by means of Flax TRM-to-masonry substrate systems shear bond tests. In addition, the issue of the durability of TRM elements was addressed as well;
- Chapter 5 deals with the assessment of the efficiency of Flax TRMs as external strengthening system of masonry structural elements to enhance their in-plane shear capacity. The results of the experimental study were discussed and compared with similar studies available in the literature, and with the model proposed by standard regulations;
- Chapter 6 proposes some observations concerning the displacement field of the TRM strengthened masonry walls subjected to diagonal compression tests, by compering such response with that exhibited by Flax TRM tested in tension;
- Chapter 7 comprises the application of innovative techniques aimed at improving the mechanical performance of Flax TRMs. The aim is to show that there is still room for improvement in the performance of plant based TRMs employed as reinforcement of structural elements, and that such enhancement of performance is easily achievable by means of available technical solutions;
- Chapter 8 shows the main outcomes and the perspectives of the research study.

1.3 Context of the Research

The study is the result of a Ph.D. scholarship awarded by the Italian Ministry for Education, University and Research (MIUR) as part of the programme "Dottorati Innovativi a caratterizzazione industriale" financed by the European Union (Structural Funding ERDF-ESF for "Research and Innovation" 2014–2020).

The Ph.D. project, carried out at the Doctoral school "Risk and Sustainability in Civil, Architecture and Environmental Engineering Systems" (Department of Civil Engineering of University of Salerno), included a collaboration with the company Engineering s.r.l. (Avellino, Italy), and with the Laboratory of Composite Materials for Construction, LMC2 (University Claude Bernard Lyon 1, France).

The collaboration with the French institution was strengthened by means of bi-nationally supervised Ph.D. thesis collaboration agreement between University of Salerno and University Claude Bernard Lyon 1, that entailed the enrolment of the Ph.D. student also to the Doctoral school "Mechanic, Energetic, Civil Engineering, Acoustic" (ED 162, MEGA) of University Claude Bernard Lyon 1 (France). Therefore, the Ph.D. research was developed at both the institution partners in the collaboration agreement.

The large experimental plan pursued in the research study, includes several investigations at different scale of analysis, and with respect to several aspects, such as mechanical, physical and chemical. It required, therefore, many instrumentations and appropriate expertise to be properly executed. In addition to the Ph.D. candidate institutions, a collaboration with external institutions was established as well. The main contribution of each institution is reported below.

Ph.D. candidate institutions:

- the Structural Engineering Testing Hall, Str.Eng.T.H. laboratory (Department of Civil Engineering of University of Salerno, Italy): a laboratory with expertise in the characterisation of the mechanical behaviour of civil engineering structures. Within the laboratory, to which the Ph.D. candidate is affiliated, the mechanical characterisation of the Flax TRM composite was carried out;
- the Laboratory of Composite Materials for Construction, LMC2 (University Claude Bernard Lyon 1, France): a laboratory with expertise in the field of composite systems for application in civil engineering structures. Within the laboratory, to which the Ph.D. candidate is affiliated, the analysis concerning the interaction between the composite and the masonry, and the structural investigation of masonry externally strengthened by Flax TRMs, were carried out;
  External institutions:
- the Industrial Engineering Department of University of Salerno, Italy: a structure with expertise in the physical, mechanical and chemical characterisation of materials. The collaboration concerned the physical and mechanical characterisation of the flax textile adopted within the study. Moreover, the institution provided the chemical equipment and expertise necessary to carry out the durability tests;
- the Centre of Sustainable Materials and Technologies, NUMATS (COPPE-Federal University of Rio de Janeiro, Brazil): a laboratory with expertise in the

study of sustainable materials and solution in the construction field. At the laboratory, within a three months research stay, the application of technologies to improve the mechanical performance of Flax TRMs, was carried out;

- the company Engineering s.r.l. (Avellino, Italy): within the company, which operates in the field of structural diagnostic of existing constructions, a six months internship provided an overview of the fields of application in which it is possible to concretely apply the results of the research.

References

1. Ambraseys N (2009) Earthquakes in the mediterranean and middle east: a multidisciplinary study of seismicity up to 1900. Cambridge University Press
2. Codispoti R, Oliveira DV, Olivito RS, Lourenço PB, Fangueiro R (2015) Mechanical performance of natural fiber-reinforced composites for the strengthening of masonry. Compos Part B: Eng 77:74–83
3. Indirli M, Kouris LAS, Formisano A, Borg RP, Mazzolani FM (2013) Seismic damage assessment of unreinforced masonry structures after the Abruzzo 2009 earthquake: the case study of the historical centers of L'Aquila and Castelvecchio Subequo. Int J Arch Herit 7(5):536–578
4. Kaplan H, Bilgin H, Yilmaz S, Binici H, Öztas A (2010) Structural damages of L'Aquila (Italy) earthquake. Nat Hazards Earth Syst Sci 10(3):499–507
5. Kouris LAS, Triantafillou TC (2018) State-of-the-art on strengthening of masonry structures with textile reinforced mortar (TRM). Constr Build Mater 188:1221–1233
6. Ramesh M, Palanikumar K, Reddy KH (2017) Plant fibre based bio-composites: sustainable and renewable green materials. Renew Sustain Energy Rev 79:558–584
7. Spiegel R, Meadows D (2010) Green building materials: a guide to product selection and specification. Wiley
8. Tolêdo Filho RD, Scrivener K, England GL, Ghavami K (2000) Durability of alkali-sensitive sisal and coconut fibres in cement mortar composites. Cem Concr Compos 22:127–143

Chapter 2
State-of-the-Art Review and Aims of the Thesis

Through the ages, the Mediterranean basin was interested by earthquakes events considered among the most destructive natural hazards, if not to the human life itself, then more likely to the works of man. The region, densely populated since ancient times, is characterised by historical buildings created using traditional construction techniques, hence not specifically designed to bear earthquake actions [3]. Most of these constructions, especially in the centres of old towns, consists of masonry structures, typically characterised by significant capacity towards gravity loads, but not always capable to resist horizontal loads applied both in-plane or out-of-plane. Typical collapse mechanisms observed as a result of catastrophic seismic events were out-of-plane failures due to poorly connected crossing walls or lack of connection of non-structural hollow clay tile walls, and in-plane failure mainly due to a lack in shear capacity, especially in presence of window or door openings [56, 57].

It is therefore essential to implement efficient strategies aimed at reducing the vulnerability of existing buildings in order to lower the seismic risk. Several retrofitting techniques were developed over the years aimed at conferring to the entire building a global response in accordance with the safety performances proposed by the standards.

The concept of seismic retrofitting is based on two main approaches: the first one aims at reducing the earthquake "demand" on the structure, the other one aims at increasing the "capacity" of the building [104]. Different retrofitting solutions may be chosen by adopting one of the two approaches, or by combining them, generally depending on the construction typology, structural scheme and the specific use of the building. With respect to the first approach, the reduction of the seismic demand of the building can be achieved by implementing measures specifically designed to optimise its structural dynamic behaviour. Building base isolation, installation of dumper devices and remodulation of the mass distribution are possible retrofitting techniques falling within this category. With respect to the second approach several solutions are nowadays available to upgrade the individual element strength: confinement/jacketing techniques; addition of thin surface treatments (shotcrete, coatings, grouted steel bars); repointing; internal reinforcement (typically by means of steel

G. Ferrara, *Flax-TRM Composite Systems for Strengthening of Masonry*, Springer Theses,
https://doi.org/10.1007/978-3-030-70273-1_2

bars inserted within masonry); tying; grout injections, filling in door or window openings. The use of these techniques, also defined as traditional retrofitting interventions, is well-established and is largely adopted for retrofitting of masonry elements. In the last decades innovative techniques were developed to enhance the strength of structural elements by using composite materials. The so-called Fibre Reinforced Polymer (FRP) composite systems, thanks to the employment of high-performance fibres, are considered a viable solution to solve, or lessen, the effects of overloading. FRP systems in the form of laminates or rods consist of long fibres embedded in polymeric matrices. High strength fibres, such as carbon, glass, aramid, PBO or basalt, are typically adopted.

The success in the use of such innovative retrofitting systems mainly lies in their tailorable performance characteristics, ease of application and low life cycle costs together with high strength-to-weight and stiffness-to-weight ratios [98]. Although resulting particularly suitable in concrete structural applications, FRP systems showed some drawbacks when applied to masonry elements. The non-permeable nature of the polymeric resins, and a strength much higher than that one of the substrates on which FRP are applied, may create a problem of compatibility between the composite and the masonry. Moreover, they are characterised by lack of ductility, susceptibility to fire, and installation problems related to the process of thermosetting adhesive being moisture and temperature dependent, and to the toxicity of the resin that requiring specialised workers [100]. To overcome such issues during the last years inorganic-matrix composite systems were developed by embedding high strength fibres in cement or lime-based mortars rather than within polymeric resins. Such innovative materials, herein defined as Textile Reinforced Mortar (TRM) composite systems, result particularly suitable as reinforcement of masonry elements, and besides, they do not present some drawbacks typical of FRPs such as low fire resistance and toxicity nature of the resins [97].

2.1 Textile Reinforced Mortar Composite Systems

Textile Reinforced Mortar composite systems, hereinafter referred to as TRMs, consist of high strength long fibres embedded in inorganic matrices, cement- or lime-based mortars, as creating laminates capable to significantly improve the strength of the structural elements on which they are applied. Several acronyms were adopted in literature to define the same kind of inorganic-matrix composite system, such as Fabric Reinforced Cementitious Matrix, FRCM; Textile Reinforced Concrete, TRC; Cementitious Matrix Grid, CMG; Inorganic Matrix Grid, IMG; Composite Reinforced Mortar, CRM [79].

TRMs represent an innovative reinforcement technique particularly suitable for applications to masonry elements, in respect of which they are preferred to the FRPs. The latter, due to a high specific strength, are often implemented in the form of strip laminates that once applied on masonry surfaces create significant concentration of stress on the substrate typically characterised by a low adhesion capacity. The use of mortar, characterised by fine aggregates, entails the necessity to use fabrics arranged with less dense meshes. As a consequence, TRMs are typically applied all over the masonry elements, allowing a more uniform distribution of the stress on the substrate. Moreover, the affinity between the material employed and the masonry, and their permeability, make TRMs more compatible to masonry substrate with respect to FRPs [53].

TRMs are characterised by good mechanical performance, reversibility and versatility of the intervention, and they offer the possibility to perform several different solutions by choosing between several types of fabric and mortar available on the market. In the light of these considerations, they are considered an effective, practical and cost-efficient solution for the seismic retrofitting and repair of structures. The use of TRMs significantly increased in recent years with numerous applications in civil engineering field as strengthening system of existing masonry elements.

In the last decade TRM systems were more and more proposed and developed as externally bonded reinforcements with several applications to cultural heritage after significant earthquake events such as "L'Aquila 2009" and "Emilia 2012" [53]. Masonry vaults were strengthened by applying TRMs in order to provide a structural stability of the element also when subjected to significant horizontal actions. The composite can be applied, below and over the vaults, in the form of strips to improve the strength of more stressed specific sections, or all over the vault surface when bidirectional oriented fabrics are adopted (Fig. 2.1).

Masonry walls were reinforced by applying all over the surface TRMs in order to improve both the in-plane and the out-of-plane capacity of the element. In some applications the upgrading intervention by means of TRM was integrated by steel plates to guarantee a correct anchorage and post-tensioning of the composite strip (Fig. 2.2).

Fig. 2.1 Extrados retrofitting of masonry vaults by applying TRM all over the surface [53]

Fig. 2.2 TRM-steel mixed intervention to upgrade a masonry element capacity [53]

Since their emergence in the market, TRMs suddenly became a very popular solution, even though there were still a lack of knowledge with respect to mechanical performance and durability of such composite systems. In order to cover this gap, and to encourage an aware use of composites in the rehabilitation of cultural heritage, several researches focused their attention on the investigation of the mechanical behaviour of TRM systems. Scientific committees were created with the aim of providing guidelines on qualification, installation, design and control, which represent crucial aspects for the effectiveness of the strengthening work.

2.1.1 TRM Characterisation Tests

Several kinds of fibres can be adopted to create TRMs by embedding them in cement, lime or geopolymer mortar matrices, possibly enriched with short fibres and/or polymeric additives. Carbon, glass, basalt and PBO fibres are commonly adopted because of their mechanical features such as high rupture stress, high stiffness, very low weight, and linear elastic behaviour up to failure. The combination of these materials and the possibility to choose the best suited geometry, make it possible to produce a large variety of TRMs depending on the need of the specific case of application. This wide choice necessarily entails a significant variability of the mechanical properties that this reinforcement technique can offer. It is necessary, therefore, to standardize the qualification procedures and to define the best characterisation methods in order to provide a valuable support for both producers and practitioners in the material selection and/or design. The scientific community focused great attention on the understanding, hence the improvement, of TRMs mechanical behaviour with the aim of providing the necessary tools for the correct designing of technical interventions by using such composite systems.

2.1.1.1 Tensile Test

Although the elements reinforced by TRMs are generally subjected to several load conditions (such as compression, bending moments, shear, etc.…), the composite itself is meant to exhibit its strength in tension. Therefore, the qualification procedure includes direct tensile tests to derive the basic material mechanical properties [5]. TRM coupons rectangular in shape, are tested in tension monotonically, in displacement control, up to the failure of the composite. The mechanical response strongly depends on the specimens' geometry, the test set-up and the monitoring technique employed to derive the mechanical parameters. A scientific technical committee, RILEM TC 232-TDT, was specifically organised with the aim of providing recommendations concerning the test method to determine the load bearing behaviour of tensile specimens made of textile reinforced concrete [85]. The document represented a valid reference both for researchers and practitioners to reproduce a standardised qualification procedure in order to obtain reliable results, comparable with similar experimental activities.

The specimens to test in tension are usually sandwich-type, performed by applying several layers of mortars, 3 ÷ 5 mm thick, alternate with textile plies. The thickness of the specimens varies between 6 and 20 mm, the width between 40 and 80 mm and the length between 400 and 600 mm. Specimens are cured in ambient conditions for a period generally considered of 28 days, due to the use of cementitious- or lime-based mortars whose characteristic mechanical properties are computed at that time [15, 59].

The specimens include two gripping areas at the edges, through which the load is applied to the specimen, followed by a transition zone in proximity to the clamping system and a measurement range in the middle (Fig. 2.3). Strains are registered by

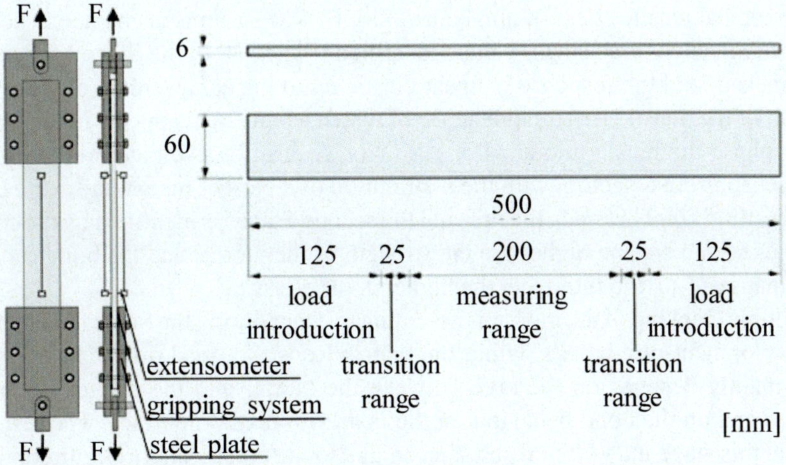

Fig. 2.3 Specimen typical geometry and test set-up [85]

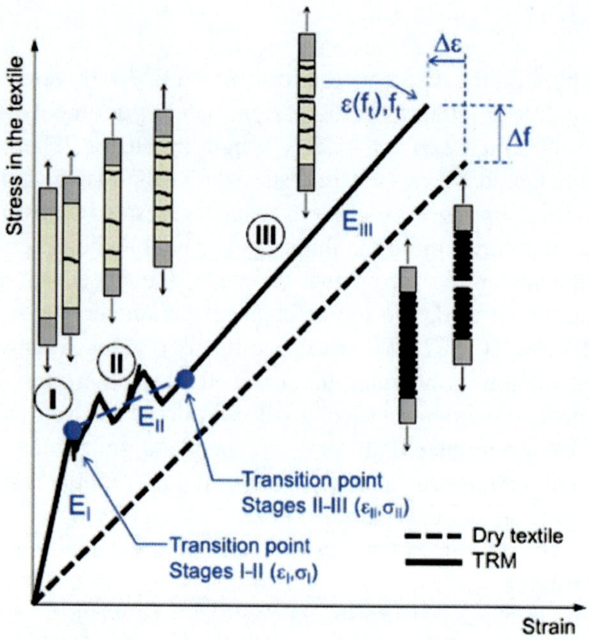

Fig. 2.4 TRM typical tensile response [35]

means of strain gauges and/or are derived by LVDT measurements. Due to developments of cracks along the free length of the specimens, displacement and strain transducers should be accurately placed along such distance, close enough to each other so that a large number of them is within the base length. The use of Digital Image Correlation (DIC) analysis, integrated with traditional devices, may provide useful information concerning the strain and displacement field of the specimen external surface, hence an accurate analysis of the crack pattern within the matrix [8].

The typical tensile stress–strain behaviour of TRM systems is characterised by a response in which three stages can be identified (Fig. 2.4). In the Stage I the mortar is uncracked, and the response is linear elastic up to the achievement of the tensile strength of the matrix in correspondence of which a first crack appears in the matrix. In this phase the tensile modulus of elasticity, E_I, can be estimated referring to a homogeneous cross section with the assumption of a perfect matrix-to-textile bond. The transition point between the first and the second Stages is identified by the couple of values σ_I and ε_I. The higher are the strength of the mortar and the bond capacity at the matrix-to-textile interface, the higher is the stress σ_I.

Multiple cracking of the mortar is assuming to occur during the Stage II. Generally, the development of a cracks within the matrix leads to a load drop whose significance mainly depends on the ratio between the mortar and the dry textile tensile stiffness and on the bond behaviour at the matrix-to-textile interface. The response curve in this stage may often result uneven due to the crack pattern occurred, therefore the stiffness E_{II} can be derived by a linear interpolation in a suitable range of data. It is important, to this purpose, that the strain and/or displacement recorded during the tests are referred to a measuring range long enough to include a sufficient number of cracks being representative of the mechanical behaviour of the system. Several aspects influence the form and the number of the cracks, such as the grid spacing of the textile, the interlocking behaviour between textile and mortar, the boundary conditions reproduced during the test, the amount of textile reinforcement, the specimen length.

Once the crack pattern has completely developed, transition point II, a nearly linear and stiffer response is observed. This phase, named as Stage III, is characterised by a widening of the existing cracks with no developments of further other ones. In the Stage III the stiffness of the composite relies on those of the dry textile and so assumes values similar to the one of the fabric alone tested in tension. In the case of a failure mode characterised by the tensile failure of the textile yarns, the bearing capacity of the composite in terms of stress assumes values close to the tensile strength of the dry textile [35].

The three stages in the TRM tensile response may be identifiable with more or less accuracy according to the volume fraction and the stiffness of the textile adopted. TRM designed with relatively low reinforcement volume ratio may end up in a mechanical response characterised by Stage III showing zero strain-hardening behaviour. Conversely, Stage III mainly governs the response of TRMs characterised by relatively high-volume fraction of a stiff textile. In the case of significantly dense fabrics combined to weak and deformable matrices, the first two stages even become hardly identifiable [34]. Considering the poor tensile strength of cementitious matrices, and the environmental conditions to which TRM systems are exposed during their service life, is it reasonable to think the composite system to work in a cracked configuration. It is, therefore, fundamental to choose a proper amount of textile reinforcement when designing a TRM system.

The failure mode concerning TRM coupons in direct tensile tests may occur due to rupture of the textile in proximity of the gripping area, due to rupture of the textile in a section included within the measurement range or due to slippage of the textile within the mortar (Fig. 2.5). The type of failure mode is strictly related

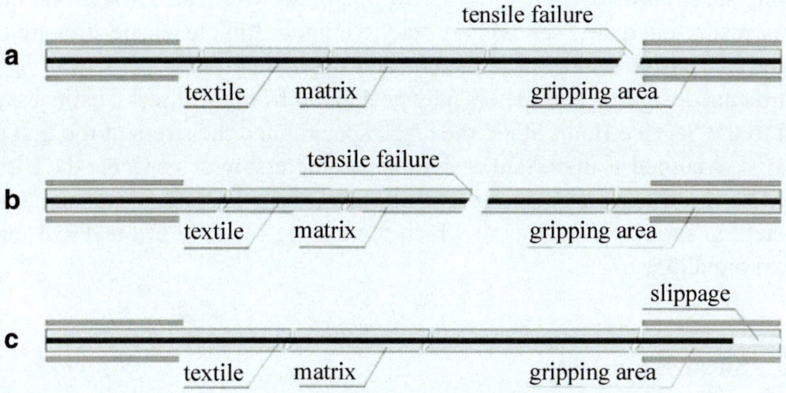

Fig. 2.5 Classification of TRM failure modes in tensile test [35]

to the boundary conditions reproduced during the test, hence the gripping system adopted. If a lateral pressure is applied on the gripping area (clamping method) the compressive stress transferred to the coupon hinder the textile to slip within the matrix in the gripping area during the test. This method induces a failure mode characterised by the rupture of the textile. If the lateral pressure is insufficient, or absent by choice (clevis method), the tensile stresses are transferred from the machine to the specimen only by means of resins glued between the steel tabs and the TRM element, and, depending on the length of the gripping area, the textile tensile strength, and the bond at the textile-to-matrix interface, the failure may occur with a slippage of the textile within the gripping area. In the case of the clamping method, in order to avoid premature failure within the transition range due to crushing of the mortar due to the lateral compression, the TRM element may be stiffened by applying FRP strips within the gripping area [4, 27].

With the aim of providing a solid database concerning the mechanical response of different TRMs subjected to direct tensile test, several research groups carried out round robin tests comparing the results deriving from different laboratories. The most common and efficient high-strength fibre textiles available in the market were considered during such tests, i.e. carbon [16], steel [36], glass [60, 87], basalt [61] and PBO [12] fibres. The tests showed in almost all the cases the typical trilinear behaviour. When glass fibres were adopted the TRM specimens exhibited a response without a clear Stage II, showing in some cases the development of a single crack. The clamping method test set up resulted the most suitable, able to guarantee almost in all the studies an entire exploitation of the fibres' strength with a failure mode characterised by textile rupture. As PBO fabrics were adopted the most common failure mode occurred consisted in a slippage of the textile within the mortar. Such behaviour highlighted a poor bond capacity with inorganic matrices of such material, with respect to its tensile strength. The increase of fibre amount obtained by overlapping more textile plies leads to higher tensile loads, however, in terms of stress a loss of efficiency may occur with a reduction of the textile exploitation due to a reduction of the bond between textile and matrix [14, 42]. The use of fabrics treated by means of coating specifically designed to improve the bond with matrix (based on the use of epoxy resins and quartz sands) may lead to higher ultimate tensile strength of the TRM and to a stiffer behaviour in the cracked stage [42].

Significant designing parameters may be derived from the direct tensile test: with respect to the Service Limit State, the crack spacing and the stress of the first crack (σ_I) may be assumed as important designing parameters; with respect to the Ultimate Limit State the Stage III stiffness, and ultimate stress and strain are fundamental parameters in such applications in which the tensile capacity of the TRM may be fully exploited [35, 87].

2.1.1.2 Shear Bond Test

TRMs are conceived as an external reinforcement of structural elements (herein masonry), hence it is necessary to consider the nature of the substrate on which they are applied to get a proper identification of the system efficiency. When the substrate gets deformed it transfers the stresses to the TRM matrix. The composite, subjected to tension, develops cracks, and in proximity of the cracks the load can be assumed to be completely borne by the textile. In such configuration, restrained by the masonry substrate, and by the stretched textile, the TRM-to-masonry system is submitted to shear loads. The shear bond test, performed on specimens characterised by a TRM strip applied on a masonry substrate, reproduces such loading conditions, and it represents, therefore, a necessary test in order to obtain an exhaustive qualification of the TRM-to-substrate system.

The assessment of the shear bond behaviour is a fundamental aspect to define, as it is the weakest mechanical phenomena, conditioning the type of failure mode, among a debonding at the TRM-to-substrate or textile-to-mortar interface bond loss and a tensile failure of the textile. It is worth emphasising that any premature bond failure, not involving the rupture of the textile, may be a solution poorly optimised since the fibres often represent the most expensive component of the reinforcing system and represents the component conferring to the composite tensile capacity [59].

Several test set-up configurations were devised in order to reproduce the actual shear bond conditions existing in real applications: single-lap or double-lap configurations, respectively for the TRM strip applied on one or both the sides of the masonry block; with one prism or two symmetric prisms, respectively if the reinforcement system is applied on one block or on two blocks symmetrically pulled in the composite direction [28, 42, 90].

Generally, the substrate is characterised by masonry prisms, but in some cases individual masonry blocks are adopted [17, 26]. Comparison between the double-lap/double-prisms test configurations and the single-lap one showed that the different test set-up led to the same mechanical response up to the peak load, but double-lap configurations failed to retain symmetric debonding on both the sides. Due to the randomness of this last aspect, and to the easier reproducibility of the test, single-shear lap test is preferred to study the bond behaviour of TRM-to-masonry systems [6].

The RILEM Technical Committee 250-CSM was created with the aim of providing recommendations to standardise the test method for the TRM-to-substrate bond characterisation [31]. The geometry of the specimen and the test set-up are shown in Fig. 2.6. The textile is partially left unbonded, clamped by using steel plates and pulled by means of the testing machine.

The specimen has to be correctly arranged so that the masonry block remain locked during the test. The textile should be perfectly in line with the TRM axis to ensure that a pure shear stress is applied to the TRM-to-substrate system. Any possible eccentricity may lead to a mixed failure mode in which both normal and shear stress develop.

The test is carried out in displacement control and the relative displacement between the reinforcement and the substrate in correspondence to the loaded end is recorded by means of LVDTs and by a steel element connected to the textile [34]. This solution proves efficiency in recording the slipping of the textile within the matrix. However, in the post peak phase, and in case of a failure mode including fibre threads rupture, the displacement data recorded may be inaccurate.

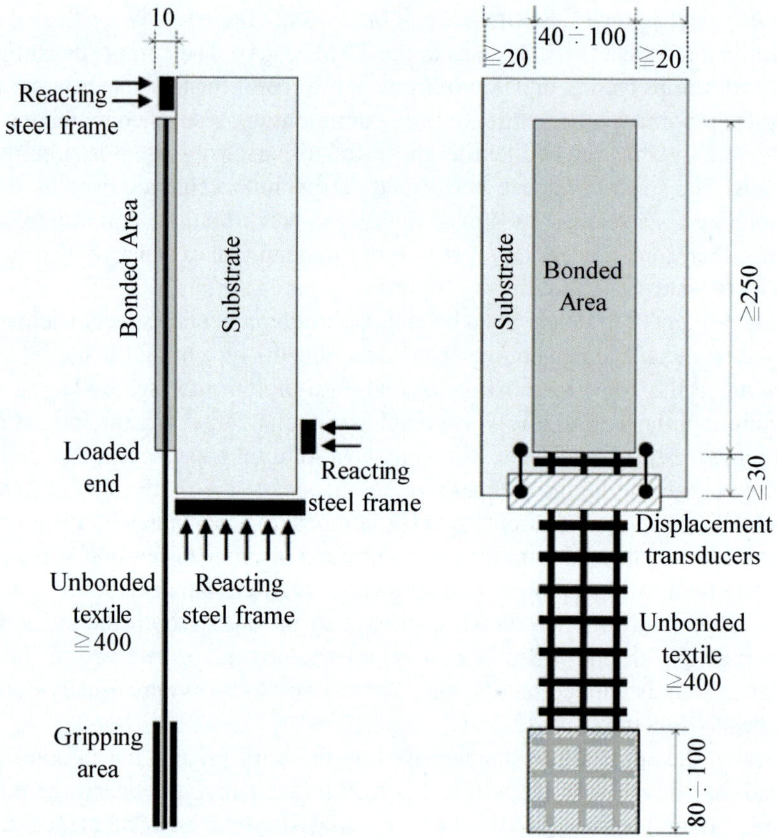

Fig. 2.6 Single-Lap Shear bond test set-up [31]

Several shear bond failure mode may be identified: loss of cohesion due to either shear failure within the substrate (mode A) or at the TRM-to-substrate interface (mode B); detachment of the textile with the outer layer of mortar (mode C); sliding of the textile within the matrix (mode D) and textile rupture either out (mode E1) or within the mortar (mode E2) (Fig. 2.7).

The cohesive failure at the TRM to substrate interface depends on the shear properties of the substrate and the adherence between mortar and masonry. The abrupt detachment of the outer layer of mortar at the textile-to-mortar interface generally occurs when dense fabrics are employed inhibiting a proper protrusion of matrix through the voids [13]. Failure modes A, B and C are characterised by a shear-slip curve with a brittle load drop, generally with short displacement values in the modes B and C (Fig. 2.8a).

Failure mode D occurs when the composite is characterised by weak textile-to-matrix bond. This aspect depends on the mechanical properties of both the mortar and the fibres, the arrangement of the fabric and the geometry of the single threads.

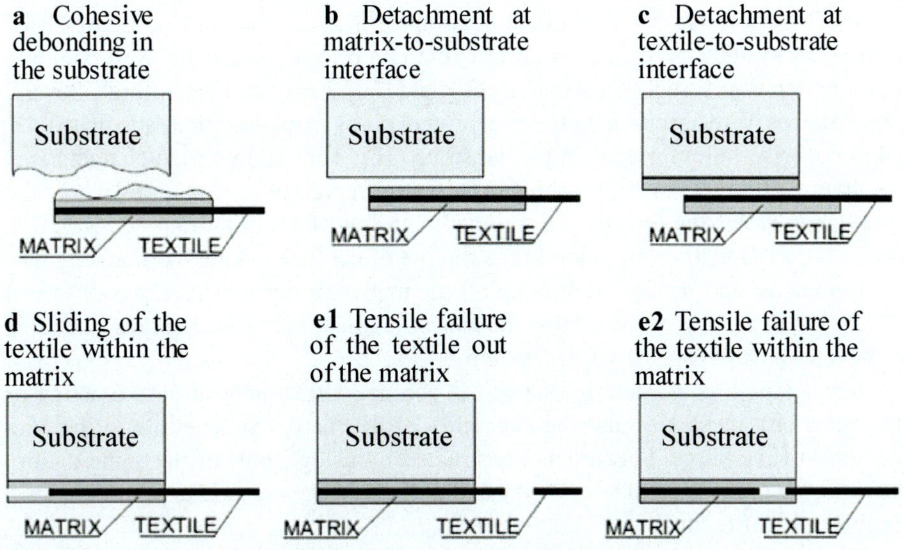

Fig. 2.7 Classification of TRM failure modes in Single-Lap Shear test [31]

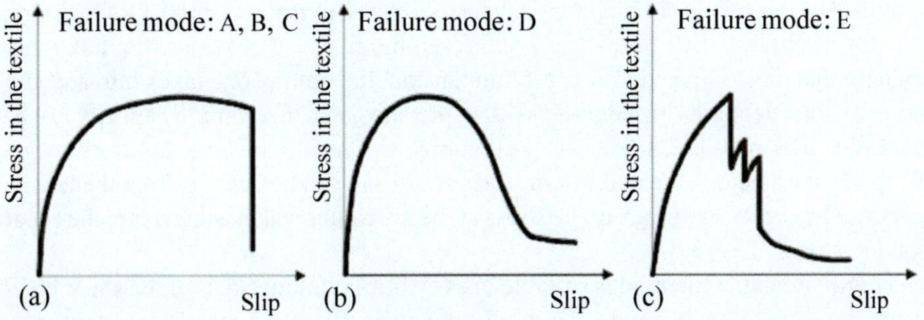

Fig. 2.8 Typical textile stress-slip curves observed in Single-Lap Shear bond tests on externally bonded TRMs related to failure modes: **a** A, B, C; **b** D; **c** E [5]

In this case the stress-slip response is characterised by a post-peak load decrease due to the progressive loss of adherence of the textile within the mortar due to the advancement of the slippage (Fig. 2.8b).

When such behaviour occurs, it is possible to improve the shear bond performance by means of fibres treatment (coating) specifically designed to improve the textile-to-mortar adherence [89].

Failure mode E may occur in case of TRM systems characterised by textiles with poor tensile strength with respect to the textile-to-mortar bond. The consequent stress-slip response is characterised by instantaneous drops of the load corresponding to the rupture of the textile bundles (Fig. 2.8c). A non-uniform distribution of the stress over the different textile threads, due—for example—to fabrics characterised by relatively sparse meshes, may contribute to the occurring of this mechanical behaviour.

Round robin tests were carried out to compare the shear bond response of TRM-to-masonry systems performed in different laboratories by using the most common high strength fibres, such as carbon, steel, glass [60], basalt and PBO fibres. Several failure modes were observed. In the case of steel fibres an exploitation ratio up to 94% led to cohesive failure modes in the substrate [16]. The failure mode characterised by slipping of the fibres within the mortar was observed in several cases where the smooth surface of the threads is combined with low strength mortars [35, 60, 61]. The use of PBO textiles often led to the rupture of the fibres along the free length of the specimens, and damage within either the matrix or the substrate was observed [12]. It was in general observed that the matrix-textile bond capacity plays a key role in the mechanical response of the system.

Specifically, it is affected by the textile geometry, treatment of yarn (coating or impregnation), mesh size and the capability of the matrix to penetrate in the core filaments of the yarns. Specimens constructed by using fibres of the same nature, but differing in these parameters, may end up in failure mode and exploitation ratio significantly different.

Whether the failure is due to loss of textile-to-matrix bond or to the textile rupture, the TRM system should be properly designed taking into account the type of failure mode in order to avoid unexpected premature collapse in the applications [44].

It is necessary, in order to get an exhaustive characterisation of the TRM system, to carry out both tensile and shear bond tests: the first one provides the maximum strength that the composite can exhibit in tension, the second one takes into account any possible debonding phenomena that prevent such maximum capacity to be achieved. It is possible, therefore, to combine the results of both the tests to get designing parameters: the maximum stress and strain can be derived from the tensile response by superiorly limiting the stress at the maximum value achieved in the shear bond test (Fig. 2.9) [59].

This qualification method is suitable in case the maximum stress of the shear bond curve corresponds to a point included in the stage III of the tensile test response.

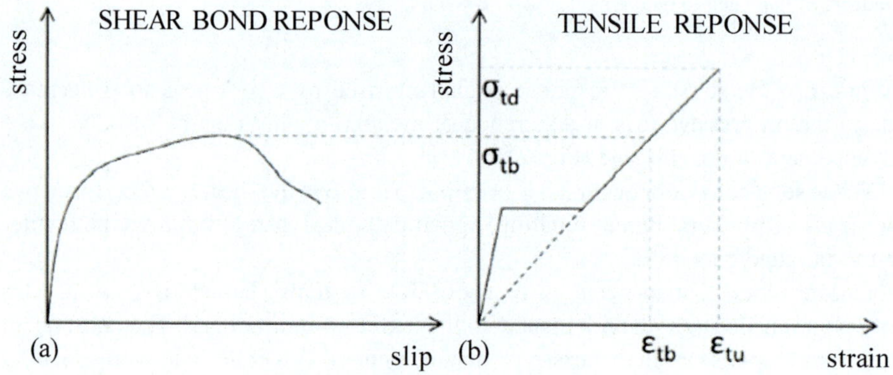

Fig. 2.9 Definition of the maximum bond stress (**a**) and strain (**b**) of TRM systems [59]

Besides, it is worth emphasising that when TRM-to-masonry specimens fail in shear bond test due to a rupture of the textile along the free length, the maximum shear stress generally coincides with the maximum one observed in tensile test.

2.1.2 Diagonal Compression Test

Tensile and shear bond represent easy reproducible tests to characterise, on a local scale, the mechanical behaviour of the composite itself and of the TRM-to-masonry system. However, it is necessary to carry out mechanical tests on larger scales of analysis in order to investigate the efficiency of the external reinforcement by TRMs on actual masonry structural elements, and to verify that qualification tests on smaller scales are enough to predict such behaviour. Depending on the load conditions to reproduce, several tests may be carried out on unreinforced and reinforced masonry elements to assess the contribution of TRMs. Masonry column were tested in compression to study the efficiency of TRM confinement [67]. In-plane capacity of masonry elements externally strengthened by TRM can be assessed by shear or bending tests [80]. Out-of-plane bending test can be carried out to reproduce the collapse that may develop as overturning, vertical flexural failure or horizontal flexural failure [55, 81].

One of the aspects of crucial importance, especially for strengthening of masonry elements toward seismic actions, is the in-plane shear capacity. Shear strength can be assessed by means of cyclic or static direct shear tests or by means of diagonal compression tests. The latter, due to the loading configuration, induces a failure mode of diagonal cracking in a stepwise form along bed and head joints or crossing masonry units. Due to the reproducibility and the reliability of the results, diagonal compression tests were often adopted to mechanically characterise the efficiency of the masonry strengthening by means of TRM systems in various types of specimens, such as two and single leaf walls, with bricks of tuff volcanic elements and by using various TRM systems differing in the geometry, nature of mortar and textile [37].

Diagonal compression (or tension/or shear) tests are typically carried out on 1.2 m × 1.2 m masonry assemblages loaded in compression along one diagonal, thus causing a diagonal tension failure with specimen splitting apart perpendicular to the direction of the load (Fig. 2.10). The test set up and procedure is described in the standard [7]. The shortening of the loaded (typically vertical) diagonal and the lengthening of the other diagonal (typical horizontal) are recorded during the test by means of specific instrumentation. On the basis of such displacements, combined with the applied load values, the maximum shear strength, strain and stiffness in the middle of the panel can be computed.

A large number of parameters strongly affect the effectiveness of TRM systems in enhancing the in-plane shear capacity masonry panels, such as geometric and mechanical properties of the matrix and mesh reinforcement, the presence of mechanical anchors, the properties of the masonry substrate. It was observed that the mechanical properties of the TRM matrix strongly affect the peak capacity of the panels.

Fig. 2.10 Clay brick wall subjected to diagonal compression test [37]

Moreover, the axial stiffness of the mesh reinforcement affects the peak capacity: when TRMs characterised by low axial stiffness textiles were adopted a smaller capacity increase was achieved with respect to those with higher axial stiffness [37]. In all cases, the reinforcing textile seemed to play a significant role in the inelastic post-peak behaviour, and even where the reinforcement was not sufficient to provide any increase of strength capacity, it seemed to enhance efficiently the displacement capacity [59].

2.1.3 Durability

Significant efforts were carried out by the scientific community to provide information about the mechanical behaviour of TRM systems and of their applications on structural elements. However, most of the study available in literature concerning such strengthening technique are pertinent with its short-term performance. Although being important to find increasingly better performing solutions, it is equally fundamental to verify that the reinforcing intervention preserve its effectiveness in time and under in-service conditions. Thus, it appears necessary to experimentally assess the long-term structural efficiency of TRM-strengthened masonry elements under various environmental exposure conditions.

Different approaches may be adopted to investigate the TRM durability performance. In some studies, the issue was addressed by ageing the reinforcing textile, exposing it to specific conditions, taking care of reproducing the environment created by the mortar [95, 101]. Micelli and Aiello showed that glass and basalt fibres are highly sensitive to alkaline environments, while steel and carbon textiles are less susceptible to that environment (Fig. 2.11). Moreover, it was highlighted that the alkalinity of the environment cannot be related to the pH value of the solution in which

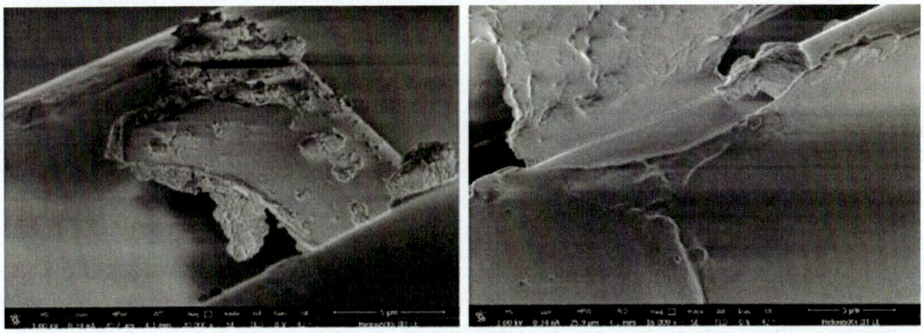

(a) (b)

Fig. 2.11 Glass fibre deterioration through **a** the surface and **b** the thickness due to exposure to alkaline environment [68]

the fibres are immersed alone, but it should be ensured that the solution contains chemical components able to reflect the real service conditions created by mortars, i.e. lime- or cement-based. The temperature represents an important parameter to reproduce accelerated ageing processes, however, further investigations are needed to accurately correlate the accelerated protocols with the actual mortar environmental conditions [68].

Another approach consists in assessing the decay of properties in the time directly on TRM coupons aged in specific environmental conditions and then subjected to direct tensile tests [86], [33]. In order to get an exhaustive evaluation of the durability performance of TRM composites, the specimens may be immersed, for a time of 1000 h, in alkaline (sodium bicarbonate solution with a pH level of 10), saline (3.5% sodium chloride solution) and water solutions at a temperature of 23 °C [72]. Moreover, in order to reflect the temperature changes to which TRMs are subjected in service conditions, freeze–thaw cycles were reproduced by alternating periods of 12 h at constant temperature of 37.7 °C to period of 4 h at −18 °C. Durability protocols represent an efficient solution to analyse the TRM ageing processes, and so to propose and verify solutions to improve the long-term efficiency of the reinforcing system. However, aggressive environments may significantly reduce the design limits and so they should be carefully considered [73]. On the other hand, ageing protocols at environmental temperature do not allow to assess the long-term durability of the composite. More elaborated protocols, perhaps by adopting higher temperatures to reproduce accelerated ageing conditions, would be needed to define the decay of the mechanical performance of TRMs in the time.

Some studies focused their attention on the efficiency of TRMs when submitted to high temperatures [10, 11, 69, 91, 96]. It was observed that textile coating treatments can improve the high-temperature performance of the system delaying the deterioration of the fibres [29]. However, such types of treatment should be specifically designed to this purpose: inappropriate coating treatments may lead to even worse performances with respect to untreated fibre based TRMs by causing a loss of adherence at the textile-to-mortar interface [41].

As well as for the short-term mechanical response, also in the case of the durability assessment it is important to analyse the entire TRM-to-masonry system to take into account any ageing phenomena involving the substrate. To this purpose masonry block specimens reinforced by TRM strips were subjected to shear bond test after being exposed to ageing protocols: such as immersion in water and salt solutions, and/or subject to wet-dry cycles. Franzoni et al. proposed an ageing protocol that produced a realistic salt accumulation in the specimens. By analysing a steel-TRM-to-masonry system, it was showed a limited effect of sulfate exposure on the stress-transfer mechanism between the composite and masonry highlighting the good efficiency of the reinforcement under the analysed exposure environment [50]. Glass-TRM-to-masonry elements showed a significant change in the mechanical response after different environmental exposures: the short term response was characterised by a failure mode with a slipping of the textile within the mortar,specimens immersed in salt solution experienced a rupture of the fibres in the free length; specimens subjected to wet-dry cycling were characterised by a debonding crisis at the TRM-to-masonry interface [40].

Although pioneering studies showed various solutions to study the durability of TRM systems, further efforts are required to better define the efficiency of the proposed ageing protocols, and to deeply understand the long-term mechanical response of the composite. To this purpose the technical committee RILEM TC IMC was recently created with the aim of clarifying scientific aspects concerning the strengthening of existing masonry structures with TRM materials, particularly focusing on those aspects related to durability and structural reliability in the long-term (RILEM TC IMC).

2.1.4 Codes and Guidelines

Ever since they appeared on the market, TRM composites immediately gained great interest from practitioners finding increased application in interventions of reinforcement of masonry structures. Therefore, it became necessary to implementat guidelines and acceptance criteria as support of the materials' qualification and of the strengthening interventions design. The aim of this paragraph is to provide an overview of the main specifications required by standards and recommendations available today concerning the use of TRMs.

The AC434, acceptance criteria for masonry and concrete strengthening using Fabric-Reinforced cementitious Matrix composites (FRCM, herein referred to as TRM), were developed in the US with aim of providing guidelines for the evaluation of such innovative strengthening method where the codes did not provide yet requirements for testing and determination of structural capacity, reliability and serviceability of these products [1]. Within the document a FRCM is defined as "a composite material consisting of a sequence of one or more layers of cement-based matrix reinforced with fibres in the form of open grid (mesh). When adhered to concrete or masonry structural members, they form an FRCM system". The document provides an overview of the mechanical tests needed to characterise TRM systems. Characterisation tests are needed on the components of the composite system, such as mortar and textile mesh. Moreover, it is suggested to carry out tensile tests to be conducted on coupons cut from TRM panels laid up using a procedure similar to that in the actual in-service applications. In addition, in order to define the bond strength, it is mentioned to carry out tensile bond tests on TRM systems applied onto substrates in accordance with the applicant's instructions. An overview of the test methods to investigate the structural performance of TRM reinforced elements is provided as well, such as wall flexural tests, wall shear tests, columns axial and flexural tests. With respect to the characterisation of the mechanical performance of TRM systems, the document provides a general comprehensive approach rather than detailed specifications. Conversely, as regard the durability aspects, ageing protocols and conditions of acceptance are described in detail. According to the document specimens should be conditioned for one week in humidity chamber, then subjected to twenty freeze–thaw cycles, each one consisting in a minimum of four hours at −18 °C, followed by 12 h at 37.7 °C. The specimens shall retain at least the 85% of the tensile and shear properties of control specimens in order to be in line with the conditions of acceptance. Environmental durability tests consist in the immersion of TRM coupons in three different solution: deionised water, salt-water, alkaline solution (with pH equal to 9.5 or higher). The three treatments should last for 1000 and 3000 h and should lead to a retainment respectively of the 85 and 80% in order to respect the conditions of acceptance [1].

The American Concrete Institute developed the ACI 549.4R-13, guide to design and construction of externally bonded FRCM systems for repair and strengthening concrete and masonry structures [2]. Within the guidelines background information and field applications concerning TRM systems are widely discussed. The main properties of TRM systems and TRM-strengthened structures are reported, and the structural design procedures are described. No novelties with respect to the TRM qualification procedures and durability requirements are introduced in this document with respect to the AC434. The guidelines provide the main tools in view of designing of interventions for the maintenance and repairing of existing structures. Specifically, formulations are proposed to evaluate the capacity to resist out-of-plane and in-plane loads of masonry elements externally strengthened by TRM systems. The section analysis is based on the following assumptions: strain compatibility between masonry and TRM composite, plane sections remain plane, the TRM is characterised by a bilinear behaviour to failure and its contribution prior to cracking is neglected. The mechanical strength parameters considered are the maximum tensile strength and strain of the TRM, and the stiffness concerning the response of the composite in tension in the cracked configuration. The approach proposed by the guidelines for the assessment of both out-of-plane flexural strength and in-plane shear strength of TRM strengthened masonry elements consists in adding up the two contributions of strength of the unreinforced masonry member and if the TRM system. The latter is calculated on the bases of the TRM strength parameters, by superiorly limiting the maximum strain to the values of 0.012 and 0.004 respectively for the case of out-of-plane and in-plane load configuration. With the aim of making easier the use of the guidelines, the ACI 549.4R-13 also provide with designing examples of TRM external reinforcement of several structural elements subjected to different load conditions [2].

With the aim of standardising the qualification test procedures of TRM and TRM-to-masonry systems scientific technical committees were created ad hoc. Recommendation of RILEM TC 232-TDT illustrates the test method to determine the load bearing behaviour of TRM tensile specimens [85]. Recommendation of RILEM TC 250-CSM defines the test method for TRM to substrate bond characterisation [31]. Both documents describe in detail the geometry of the specimens, the test condition and equipment, the test procedure and the evaluation of test results.

With respect to the Italian legislative context, the technical code for constructions [74] requires that all materials and products adopted in constructions, when used with a structural function, should be identified, qualified and controlled. To this purpose, guidelines for the identification, qualification and acceptance criteria of TRM to be used for retrofitting of existing structures, were recently developed [62]. An interesting aspect present in the latter concerns the application of the reinforcement system. As known, the mechanical efficiency of TRMs is strictly related to a proper implementation in situ. Therefore, it is necessary that the manufacturer provides technical handbook explaining the installation procedure, specifying the range in which the thickness of the composite may vary and possible needed preparatory treatments of the substrate. Such instructions, to be respected during the reinforcement application, also represent the reference for control inspections. Also in Italy instructions for the design, execution and control of reinforcement interventions by means of TRMs were developed, provided by the Italian National Research Institute [22]. Similar to what is proposed by the American guidelines, the document includes an overview of the possible applications of the retrofitting system, the basic principles for conception of retrofitting interventions and designing formulas. The qualification procedure provides for tensile test of the textile, tensile test of TRM coupons and shear bond test of TRM-to-substrate systems. The mechanical parameters deriving from the characterisation tests, and to consider in designing formulas, are defined as follow:

- σ_{uf}, ε_{uf} and E_f, respectively representing the maximum tensile stress and strain and the tensile stiffness of the dry textile;
- $f_{c,mat}$, representing the compression strength of the matrix;
- σ_u, ε_u, respectively representing the maximum tensile stress and strain of the TRM deriving from direct tensile test (clamped configuration);
- E_I, representing the stiffness in tension of the TRM with respect to the Stage I (linear elastic phase in the uncracked configuration);
- $\sigma_{lim,conv}$, $\varepsilon_{lim,conv}$, defined as conventional limit stress and strain.

The normal stresses are conventionally referred to the cross section of the dry textile in the direction of loading. The conventional limit stress, $\sigma_{lim,conv}$, represents the maximum strength of a TRM-to-substrate system derived from the shear bond test. The conventional limit strain, $\varepsilon_{lim,conv}$, represents the deformation corresponding to the conventional limit stress in the tensile stress–strain response of the dry textile (Fig. 2.12).

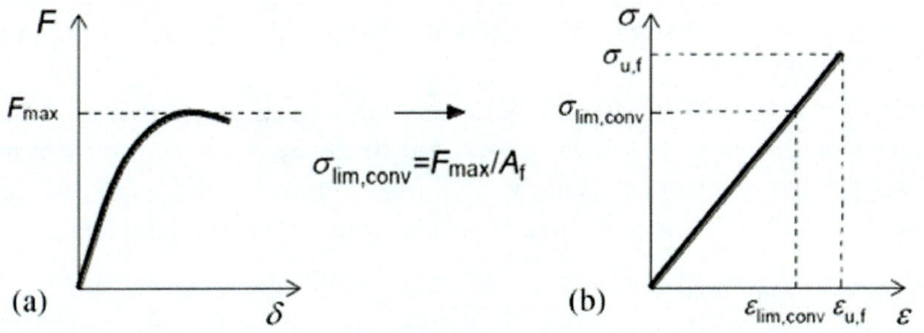

Fig. 2.12 Conventional limit stress and strain assessment by means of: **a** TRM-to-substrate shear bond test; and **b** dry textile tensile test responses (CNR-DT 215, 2018)

2.2 Use of Plant Fibres in TRM Composites

During the last decades, the sensitivity of the public opinion toward environmental issues has significantly increased. The first definition of "sustainable development" was given by the Brundtland Report "Our Common Future" in 1987, in which it is highlighted that "*Humanity has the ability to make development sustainable to ensure that it meets the needs of the present without compromising the ability of future generations to meet their own needs*" [99]. Such concept implies the improvement of technology and social organization to make way for a new era of economic growth in line with the current elaborated model of circular economy, defined as "*a regenerative system in which resource input and waste, emission, and energy leakage are minimised by slowing, closing, and narrowing material and energy loops*" [51]. Moreover, the need of more sustainable solutions is recognised as one of the main "societal challenges" of our days also by the EU Research and Innovation program "Horizon 2020". Therefore, limiting the environmental "impact" of human activities is nowadays considered as a priority in all industrial fields. Among others, the construction sector needs to get "greener" by reducing both the demand of energy and raw materials, and the emission of green-house gases [66]. Moving in this direction new sustainable building materials are conceived by promoting the reuse of waste-materials and the use of the so-called eco-friendly raw material sources characterised by a low environmental impact [24].

Composite materials lend themselves well to the implementation of several solutions aimed at obtaining more sustainable systems. For instance, the adoption of natural fibres (also called bio-fibres or cellulosic fibres and hereinafter referred to as plant fibres), in lieu of industrial ones, as reinforcement of both organic and inorganic matrices, is one of the possible features of a new class of materials generally referred to as biocomposites. In recent years, such use experienced a considerable increase in engineering applications. In 2003, the use of plant fibres in composites in EU, was estimated to be around 43,000 tonnes [63]. In 2010, the amount increased to around 315,000 tonnes, representing about the 13% of all types of reinforcing fibres available (carbon, glass, etc.…) and estimated to be around 28% by 2020 [18]. The US Departments of Agriculture and Energy set goal of having at least 50% of all basic chemical construction blocks created from renewable and/or plant-based green sources by 2050 [70].

It is reasonable to consider bio-composite as the next generation of structural materials. For this reason, several efforts are made by the scientific community to investigate the mechanical behaviour of plant-fibres-based composites, and to improve it according to the performance levels required by structural applications.

2.2.1 Plant Fibres

Based on their origin, natural fibres can be classified as plant, animal or mineral [103]. Plant fibres, in turn, are subdivided into wood fibres and non-wood fibres. The latter include different kinds of fibres depending on the part of the plant they are taken out: leaf (sisal, pineapple, àbaca), bast (hemp, flax, jute, kenaf, ramie), seed (cotton), fruit (coir, kapok), straw (rice, corn), grass (bamboo). The great availability in many countries, their cost-effectiveness and low density, are some of the advantages that led to a great interest in the use of such fibres (Table 2.1). Due to their nature, they result to be biodegradable, renewable, non-hazardous and non-abrasive.

Moreover, they proved to have good mechanical properties, sometime comparable to those of synthetic fibres. However, a successful commercialization of such fibres

Table 2.1 Estimation of the plant fibres global production [102]

Fibre type	Production per year (Million tonnes)	Main producer countries
Abaca	0.10	Philippines, Equator
Cotton	25	China, USA, India, Pakistan
Coir	0.45	India, Sri Lanka
Flax[a]	0.50–1.5	China, France, Belgium, Ukraine
Hemp[b]	0.10	China
Henequen	0.03	Mexico
Jute	2.5	India, Bangladesh
Kenaf	0.45	China, India, Thailand
Ramie	0.15	China
Silk	0.10	China, India
Sisal	0.30	Brazil, China, Tanzania, Kenya

[a]The real production of flax was underestimated because the production of Canada is not considered

[b]China has announced plan to substantially increase the hemp production for textiles in the coming years to 1.5 million tonnes of fibres per year

is prevented by some issues still unsolved. The lack of data related to the durability creates a widespread scepticism that prevents the implementation of these materials in engineering areas. In addition, significant fibre contents are needed in order to have adequate reinforcement effects, with possible difficulties in mortar mixing due to reduction in workability. Another drawback is related to the reliability of the mechanical properties of plant fibres, frequently characterised by a significant scattering due to their dependence on random variables such as chemical composition, filaments geometry (diameter and length) and surface roughness [102].

Plant fibres (also named as cellulose/lignocellulose fibres) may themselves be considered as a composite whose basic chemical components are cellulose, hemicellulose and lignin. The cellulose represents the stiffest and strongest component, and its amount affects the fibre strength, stiffness and stability. Plant fibres can be classified with respect to the chemical composition, and so the proportion among the different components: i.e. cotton, hemp, flax, sisal, ramie, with an amount of about 70% of cellulose of the entire chemical composition, are considered as fibres reach in cellulose. Several other aspects may also affect the chemical composition, such as geographical and climatic conditions, the variety of the plant, soil quality, level of plant maturity and the fibre manufactory process. The combination of these features determines the mechanical properties of the fibres [58].

Depending on their nature, plant fibres may present different structures. In general, they are arranged as bundles of elementary fibres, each one characterised by a multilayer cylindrical structure. In Fig. 2.13 the typical structure of a flax fibre is

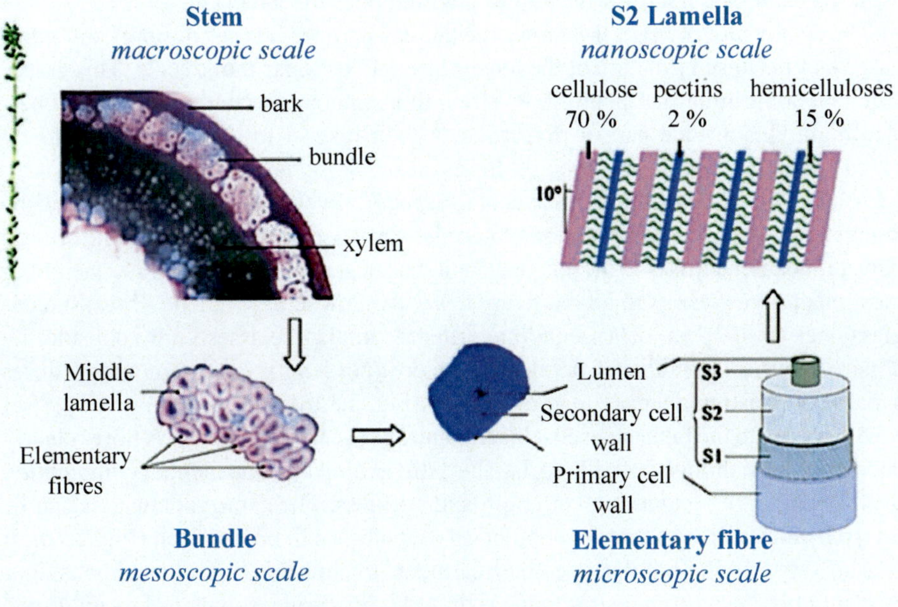

Fig. 2.13 Schematic multi-scale flax fibre structure [103]

represented from the stem (macroscopic scale) to the elementary fibre components (nanoscopic scale).

The stem is characterised by several layers. Bundles of fibres are placed right under the bark. The cross section of a bundle contains between 10 and 40 elementary fibres. At the microscopic scale the elementary fibre is characterised by concentric cell walls, differing each other in terms of thickness and chemical components. The centre part, with a small open channel in the middle, is called lumen, and it is the part responsible of water uptake. On the outer side, the primary cell wall is located around the secondary wall. The latter, with a high content of cellulose, is responsible for the strength of the fibre [102].

Physical properties also play an important role in the mechanical performance of plant fibres. Generally, the tensile strength decreases when the fibre length increases in line with the principle according to which the longer the fibre, the higher the probability of presenting imperfections leading to premature failures compared to shorter fibres [103]. The main mechanical properties of cellulosic fibres, compared with those of synthetic fibres such as carbon, aramid and glass, are shown in Table 2.2.

Flax, ramie and hemp represent the most performant fibres, whose strength and stiffness ranges of values are relatively close to those of synthetic fibres, such as glass filaments. Nevertheless, the same cannot be said with respect to the strain at failure. Such discrepancy is due to the mechanical response in tension of plant fibres, that, unlike synthetic fibres, shows a first branch that is less stiff prior to achieve a linear behaviour. Such behaviour, typical of plant fibres, consists of an elasto-visco-plastic deformation due to the re-arrangement of the amorphous parts of the thickest cell wall, the cellulosic microfibrils, aligning with the tensile axis [21].

The Ashby plot in Fig. 2.14 shows the density of several types of fibres and fibre-based composites in function of the tensile strength on logarithmic scale. This graph, although confirming that plant fibres strength is comparable to those of glass ones, also highlights a wide gap of performance with respect to carbon and polymeric based composites with high strength fibres.

As known, in addition to mechanical aspects, several other features affect the interest in the use of a construction materials. Among them surely economic criteria play a fundamental role. With this respect it was observed that the cost per weight of most plant fibres results to be relatively lower than those of synthetic fibres, such as glass ones (Fig. 2.15a). The economic criteria can also be assessed by considering a more significant parameter defined as the cost per length for an amount of fibres capable of resisting a tensile load of 100 kN (Fig. 2.15b).

According to the latter such cost discrepancy between plant and glass fibres clearly decreases, even though such fibres, i.e. flax, still exhibit a good potential combination of low cost, light weight, high strength and stiffness. The same conclusions can be drawn by considering a greater population of materials of comparison (Fig. 2.16). It is necessary to point out that specific treatments to adjust the mechanical behaviour of plant fibre-based composites to the required performances may lead to additional costs related to their use. In addition, being plant fibres materials a relatively young market concerning their supply and use, great uncertainty exists concerning the

Table 2.2 Mechanical properties of main plant and synthetic fibres [103]

Fibre type	Relative density (g/cm³)	Tensile strength (MPa)	Elastic modulus (GPa)	Specific modulus (GPaxcm³/g)	Elongation at failure (%)
Abaca	1.5	400–980	62–20	9	1.0–10
Alfa	0.89	35	22	25	5.8
Bagasse	1.25	222–290	14–27.1	18	1.1
Bamboo	0.6–1.1	140–800	11–32	25	2.5–3.7
Banana	1.35	500	12	9	1.5–9
Coir	1.15–1.46	95–230	2.8–6	4	15–51.4
Cotton	1.5–1.6	287–800	5.5–12.6	6	3–10
Curaua	1.4	87–1150	11.8–96	39	1.3–4.9
Flax	1.4–1.5	343–2000	27.6–103	45	1.2–3.3
Hemp	1.4–1.5	270–900	23.5–90	40	1–3.5
Henequen	1.2	430–570	10.1–16.3	11	3.7–5.9
Isora	1.2–1.3	500–600	–	–	5–6
Jute	1.3–1.49	320–800	30	30	1–1.8
Kenaf	1.4	223–930	14.5–53	24	1.5–2.7
Piassava	1.4	134–143	1.07–4.59	2	7.8–21.9
Palf	0.8–1.6	180–1627	1.44–82.5	35	1.6–14.5
Ramie	1.0–1.55	400–1000	24.5–128	60	1.2–4.0
Sisal	1.33–1.5	363–700	9.0–38	17	2.0–7.0
Aramid	1.4	3000–3150	63–67	46.4	3.3–3.7
Carbon	1.4	4000	200–240	157	1.4–1.8
E-glass	2.5	1000–3500	70–76	29	0.5
S-glass	2.5	4570	86	34.4	2.8

future price of these materials. For these reasons, further investigations are needed to accurately define the economic aspect.

A more comprehensive analysis aimed at investigating the potential use of plant fibres in construction materials, cannot neglect the environmental impact, an issue of primary importance nowadays. Plant fibres are known to be characterised by relatively low embodied energy, representing, the latter, the energy consumed by all processes associated with the production of a construction material (or of a building) from the mining and processing of natural resources, to manufacturing and transport. By putting in relationship such a parameter with the tensile strength, over a wide population of different types of fibres and fibre-based composites, an exhaustive comparison can be provided (Fig. 2.17).

In absolute terms, the embodied production energy concerning synthetic fibres is approximately 10 times larger than that of plant fibres, and it is highlighted that synthetic fibre-based composites require around five times more energy of production

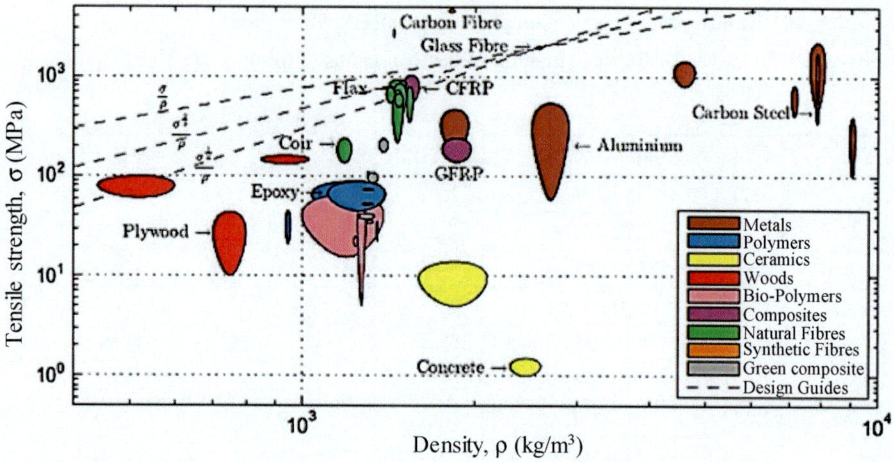

Fig. 2.14 Density—tensile strength relationship of different types of fibres and fibre-based composites [38]

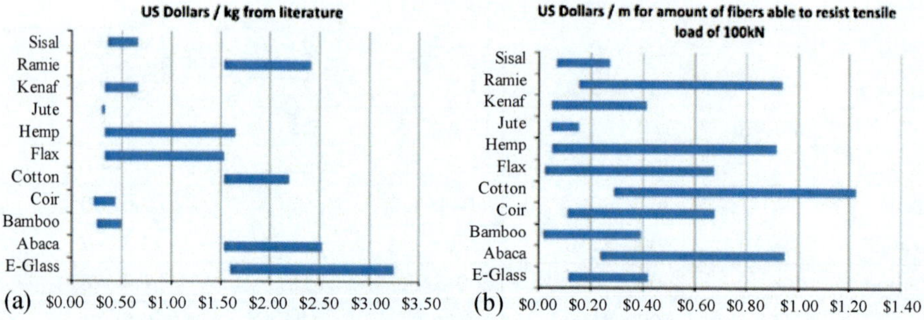

Fig. 2.15 Comparison between E-glass and plant fibres: **a** cost per weight; **b** cost per unit length capable of resisting 100 kN load [39]

than green composites [38]. By taking into account the several aspects discussed above, such as physical, mechanical, economic and ecologic ones, among the various plant fibres, flax, hemp, cotton, sisal and jute concretely represent potential candidates as reinforcement of innovative composites in civil engineering applications [103].

2.2.2 *Plant Fibres as Reinforcement in TRMs*

The use of plant fibres as reinforcement in inorganic matrices represents a smart solution to employ the potential of the fibres in TRMs, and to improve the sustainability of such an innovative retrofitting technique. Several research studies carried out to investigate the efficiency of plant-based TRMs, showed it as a valuable and

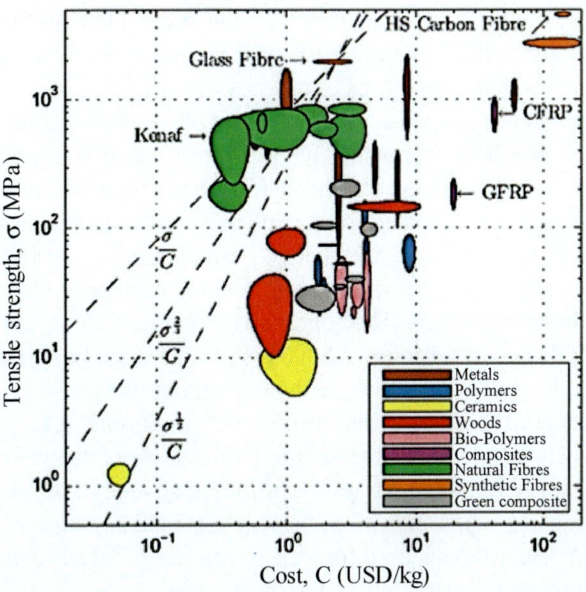

Fig. 2.16 Cost—tensile strength relationship of different types of fibres and fibre-based composites [38]

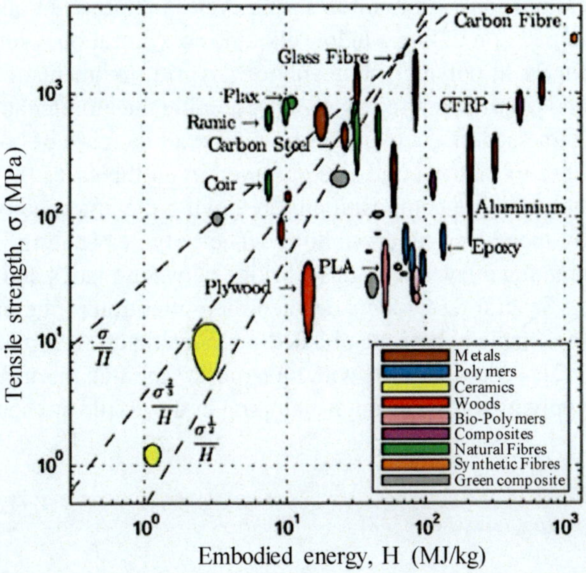

Fig. 2.17 Embodied energy—tensile strength relationship of different types of fibres and fibre-based composites [38]

promising solution [83]. Due to their nature, physical structure and mechanical properties, plant fibres are characterised by a mechanical performance different from that exhibited by synthetic fibres. Therefore, the scientific outcomes deriving from the several studies carried out in the last decade on the use of TRMs, mainly conducted by using high strength fibres, cannot be directly extended to plant fibre based TRMs. For this reason, many researchers focused their attention on the mechanical behaviour of lime/cement-based matrices reinforced by plant fibre textiles, often referred to as Natural TRMs, with the aim of assessing their efficiency. Several aspects were investigated on different scales of analysis from the fibre-to-matrix bond, to the application of Natural TRMs as reinforcement of masonry elements.

The bond behaviour at the fibre-to-matrix interface represents a fundamental aspect that affects the mechanical behaviour of the entire TRM composite and its performance as reinforcement of structural elements. Pull-out test, consisting in pulling a textile yarn embedded in a mortar cylinder (Fig. 2.18), provides shear stress vs slip (also named bond-slip) curves representative of the bond capacity between the two elements. The bond-slip law is typically characterised by a first linear branch followed by a frictional bond phase. The maximum shear bond stress, and the magnitude of the residual force depend on the materials' properties and on the embedded length. The shear-bond response of jute fibres in a cementitious matrix was studied by [48], in which a debonding failure showed that the critical transfer length is larger than 10 mm. The study also proposed an analytical model providing a bilinear bond-slip law representative of the experimental evidences. The bond between plant fibres and matrix can be improved by means of treatments applied on the fibres before embedding them within the mortar [49].

Different treatments may be applied to the fibres to improve their strength: hornification, consisting in wet-drying cycles in hot water; alkaline treatment, consisting in immersion of fibres in low alkaline concentration solutions for 50 min and in drying for 24 h; polymer impregnation; hybrid treatments, combining fibres hornification and polymer impregnation. By adopting the aforementioned treatments to sisal fibres immersed in a cement-based matrix, an increase of both fibres' tensile strength and fibre-to-matrix bond was observed in all the cases [47]. The alkali treatment modifies the fibre composition by removing the amorphous constituents of the fibre and by increasing its crystallinity. The polymer coating increases the interfacial bond creating a layer acting as a bridge between matrix and fibre.

Several researchers devoted their attention to the mechanical characterisation of the tensile behaviour of TRM coupons made with plant fibre textiles. Ghiassi et al. [52] produced TRMs with lime mortar and flax fabric and induced tensile stresses by applying the load trough clamping of the textile on the two sides of the specimen. By

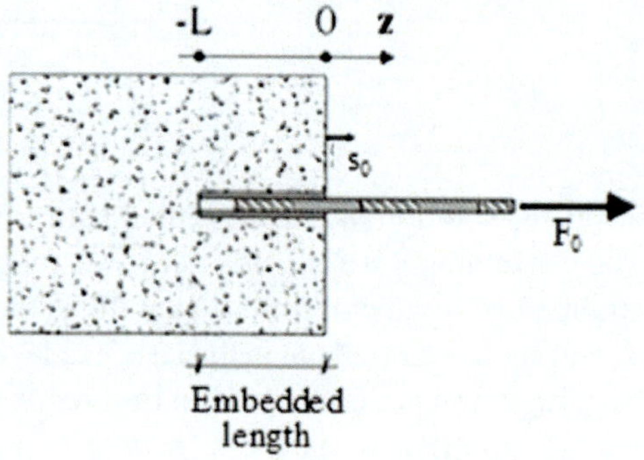

Fig. 2.18 Single fibre pull-out test scheme [48]

comparing the results with the tensile response of the dry textile, although a decrease of stiffness was observed, a significant increase of the tensile strength was evidenced, due to the increased integrity of the system provided by the mortar presence. Codispoti et al. [23] carried out tensile test on TRMs performed by using several types of textile, such as jute, sisal, coir, flax, hemp (Fig. 2.19). The different types of Natural TRM exhibited a promising mechanical response with a good exploitation rate of the entire dry textile strength. However, it was emphasised that it is fundamental to adopt appropriate thickness of the mortar, and hence appropriate volume fraction of fibre, in order to produce a mechanically efficient system.

The tensile response of Natural TRMs is characterised by the typical three phases generally exhibited by the conventional synthetic fibre-based system [75]. Unlike the latter, Natural TRMs are characterised by a second stage in which, in correspondence to the occurrence of a crack, a significant drop of the load is observed due to a discrepancy in the axial stiffness of the textile ($A_f \times E_f$) and the axial stiffness of the mortar (Fig. 2.20).

The fibre volume fraction plays a key role in the mechanical response of Natural TRMs [43]. Figure 2.21 shows the tensile strength of plant fibre composites (assessed referring to the entire cross-section area of the coupons), performed by using flax and sisal, as a function of the fibre volume fraction. It can be noted that starting from a volumetric ratio of about the 3%, a significant enhancement of the mechanical performance was experienced [19].

Natural TRM tested in tension typically exhibited a failure characterised by the rupture of the fibres. However, depending on the test set up, a slippage of the fibres within the clamped zone may occur [30].

The influence of fibre treatments, i.e. coating, on the mechanical response of several types of Natural TRMs, performed by using flax, hemp, sisal and cotton fabrics, was investigated as well. Mercedes et al. employed two different treatments by coating the fabrics either with epoxy resin or with polyester. In both cases,

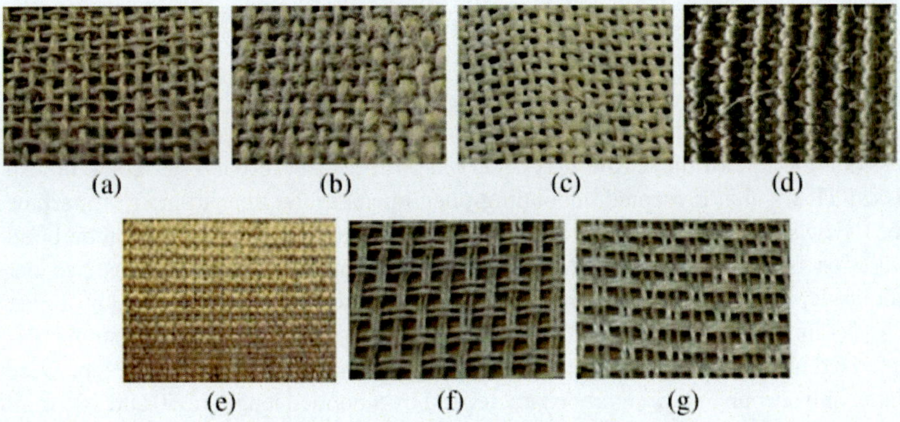

Fig. 2.19 Different types of plant fibre fabrics: jute (**a**, **b** and **c**); sisal (**d** and **e**); hemp (**f**); flax (**g**) [23]

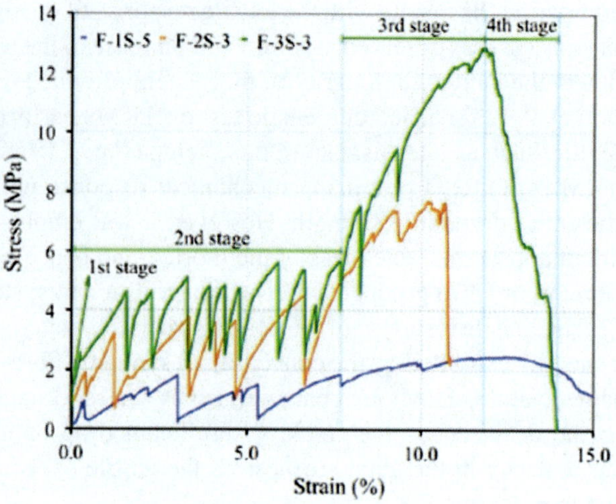

Fig. 2.20 Stress–strain response of composite specimens reinforced with one, two and three layers of flax fabrics [19]

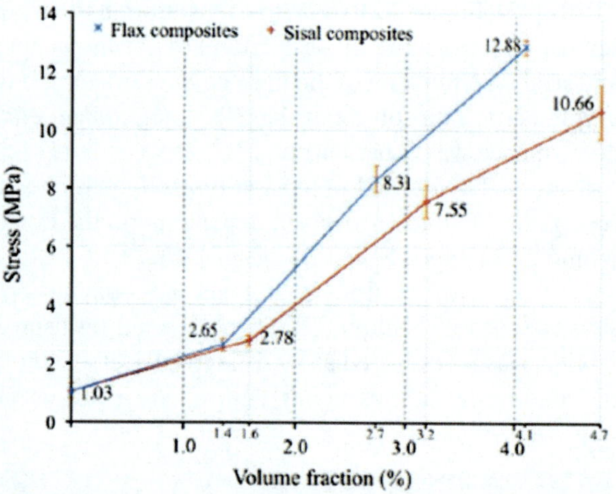

Fig. 2.21 Effect of the fibre volume fraction on tensile strength of Natural TRMs [19]

an increase of the tensile response of the textile was observed. In regards to the tensile strength of Natural TRMs, when epoxy-coated meshes were adopted, a larger exploitation rate of the textile was observed with respect to polyester-coated mesh based TRMs, that presented debonding phenomena at the fibre-to-matrix interface [65]. Few studies focused their attention on the experimental investigation on bond adhesion between Natural TRMs and masonry support. Olivito et al. carried out double-lap shear bond tests on masonry clay bricks externally strengthened by a Flax TRM composite and compared the mechanical behaviour with a similar system characterised by conventional high strength fibres (PBO) [76]. Two configurations were taken into account, respectively characterised by a bonded length of 50 and 100 mm. The experimental results highlighted clear differences between the two reinforcing systems adopted: the flax-based system exhibited a failure mode characterised by the

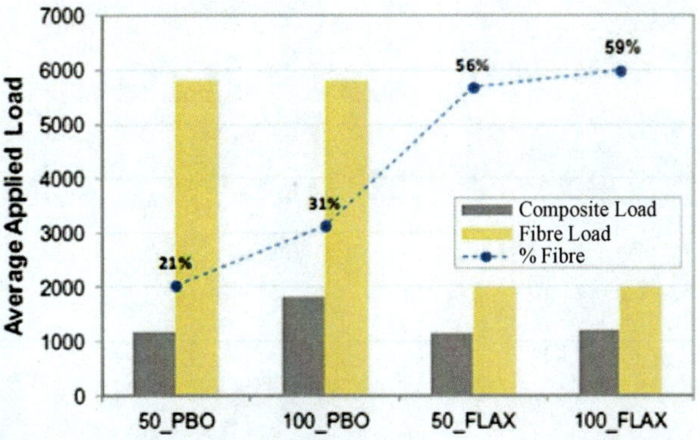

Fig. 2.22 Fibres exploitation ratio in PBO-TRMs and Flax-TRMs [76]

rupture of the textile yarns in the unbonded length, whereas PBO-based systems experienced a brittle cohesive failure mode. Flax fibres exploited their potential strength to a high degree during the test achieving almost 60% of the dry fibre tensile strength, unlike PBO textiles that used only between the 20% and the 30% of their strength (Fig. 2.22). Such experimental evidences point out the good collaboration between flax fibres and inorganic matrix, in some cases showing even more compatibility than synthetic fibres. It is important to notice that in case of TRM strengthened masonry elements under cyclic/dynamic loading, a controlled debonding of the textile within the matrix would be more desirable in terms of energy dissipation. Still it is important emphasising the possibility to increase the capacity of the system by increasing the textile amount, due to the good bond behaviour between plant fibres and mortar.

de Carvalho Bello et al. [30], carried out single lap shear tests on Sisal TRM-to-masonry systems according to RILEM recommendations [31]. Sisal fibres were impregnated by an adhesive promoter prior to the inclusion in the inorganic matrix. The failure mode observed was characterised by the tensile failure of the yarns along the free (unbonded) length, and no slip was observed at masonry-to-mortar interface. Due to an uneven distribution of the stresses among the sisal yarns, the maximum tensile strength attained was about 30% lower than the potential strength of the dry textile. Some applications of Natural TRMs as external reinforcement of masonry elements are present in literature [45]. Cevallos et al. assessed the efficiency of a Flax-TRM system for strengthening of masonry structural elements subjected to eccentric loads (Fig. 2.23). The mechanical response of different sized specimens was compared to those of elements externally strengthened by PBO-TRM composites. The reinforcement conferred to the masonry elements a substantial gain in strength and deformability. By comparing the plant fibre-based system with the synthetic based one, it was highlighted that the higher ductility of flax fibres allowed for a higher strain in the TRM forming cracks in the mortar. Such behaviour resulted in a more ductile behaviour, unlike the case of PBO TRM-strengthened elements, in which the lower strain capacity of the fibres led to debonding of the strengthening system with a consequent brittle collapse of the element [20].

Diagonal compression test represents un efficient experimental equipment for the assessment of the contribution of Natural TRMs to the in-plane shear capacity of

(a) (b) (c)

Fig. 2.23 Application of the Flax-TRM system on masonry structural elements: **a** application of the first layer of mortar; **b** embedment of the fabric into the matrix; **c** smoothing of the outer layer of mortar [20]

masonry elements [46]. Menna et al. investigated the in-plane response of either clay bricks or tuff masonry elements externally strengthened, uniformly all over the surface, by a TRM system comprising a hemp textile, properly impregnated in a flexible epoxy resin, and either a pozzolanic or a lime-based mortar [64]. The experimental outcomes showed enhanced mechanical properties in all tested panels, with a maximum shear strength increased by a factor of about 2–5. The failure occurred with a mostly uniform crack pattern that developed until the entire exploitation of the textile that achieved its tensile strength in the most stressed points (Fig. 2.24).

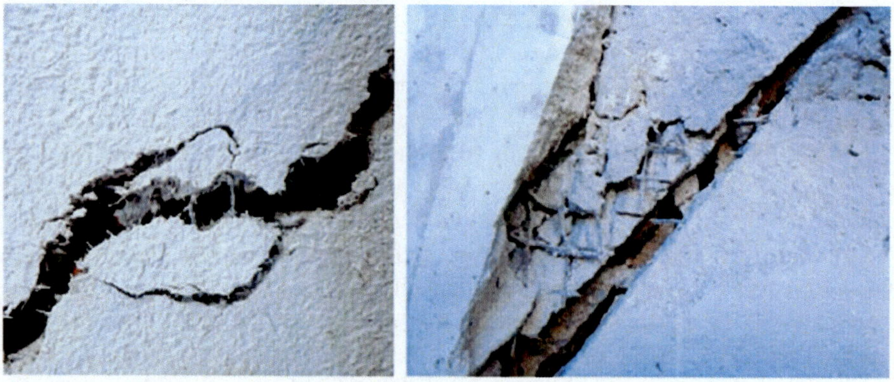

(a) (b)

Fig. 2.24 Tensile failure of textile in proximity of cracks in Hemp-TRM externally strengthened clay brick **a** and tuff **b** masonry elements [64]

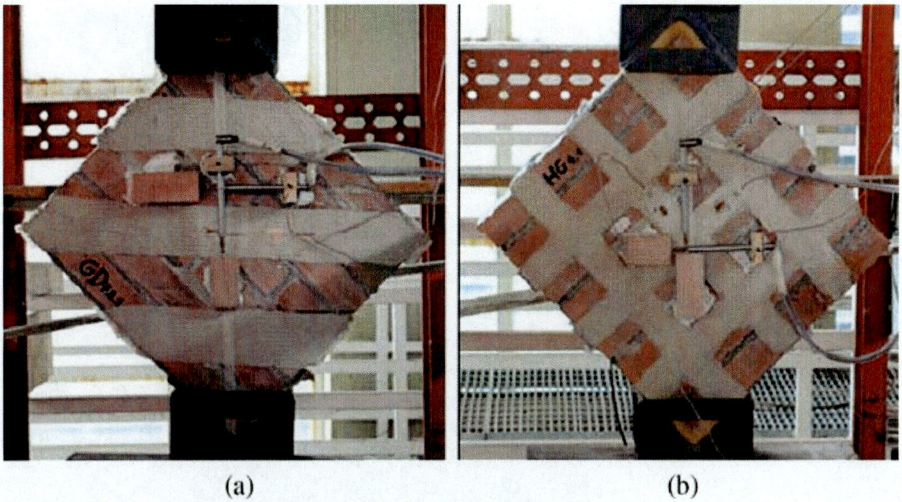

Fig. 2.25 Clay brick walls eternally strengthened by Natural TRMs arranged in strips oriented in the tensile direction (**a**); and applied in the form of bidirectional grids (**b**) [77]

Different configurations of reinforcement, with either Flax-TRM and Hemp-TRM unidirectional strips applied on the surfaces of clay brick walls, were also investigated (Fig. 2.25). An increase of the strength was observed in the range of the 20% and 40% with respect to unreinforced masonry walls tested in diagonal compression [77].

In order to define the effectiveness of plant fibre-based cementitious composites, it is necessary, in addition to mechanical investigations, to focus the attention on durability aspects with the aim of assessing the ability of the system to resist both external (weather conditions) and internal damage (compatibility between matrix and fibres). Plant fibres embedded in Portland cement undergo a degradation process due to the high alkaline environment [54]. Such degradation is due to the alkaline pore water that dissolves the lignin and hemicellulose in the middle lamellae of the fibres, weakening the link between the individual fibre structure elements. An additional phenomenon consists in the alkaline hydrolysis of cellulose, that causes degradation of the molecular chains with a consequent decrease of polymerisation of the fibre structure. Tolêdo Filho et al. [92] investigated the degradation of sisal and coconut fibres by immersing them in alkaline solutions [93]. It was observed that in sodium hydroxide (NaOH) solutions between the 30 and 40% of the strength was lost after 420 days of exposure. Even a more aggressive exposure was that of a calcium hydroxide ($Ca(OH)_2$) solution in which a complete loss of strength was evidenced after 300 days of immersion. The heavier attack of the calcium hydroxide may be attributed to the crystallization of lime in the fibre's pores [82].

Another aspect that contributes to the degradation of plant fibres in inorganic matrix composites is related to the capacity of such fibres to absorb water [94].The latter may lead to volume changes that can induce cracks within the matrix, as in the case of hygral fatigue (Fig. 2.26).

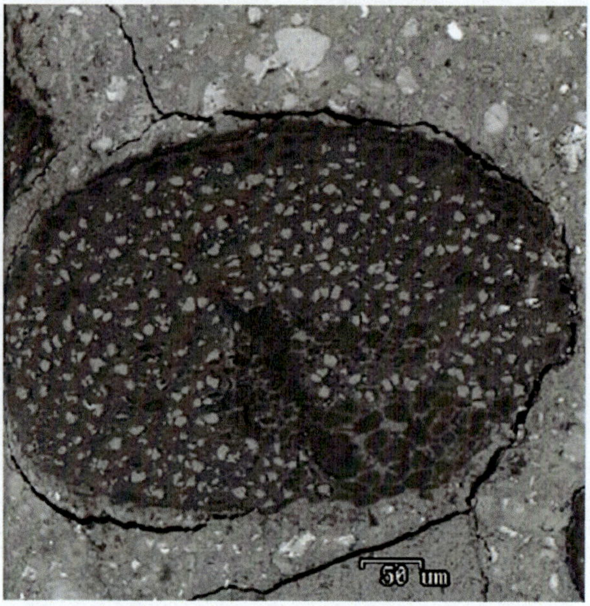

Fig. 2.26 Coconut fibre-cementitious matrix interface after 25 cycles of wetting and drying [93]

In order to improve the durability of plant fibres in inorganic matrices two approaches can be adopted: matrix modification and fibre modification [78]. With respect to the first strategy using low alkaline matrices can reduce their aggressivity, for instance by adding pozzolanic products, rice husk or fly ashes [32]. D'Almeida et al. [25], showed that by replacing part of the Portland cement by metakaolin it is possible to obtain a matrix totally free of calcium hydroxide, hence avoiding a damage of the fibre lumen, middle lamella and cell walls. As regards fibre treatments to improve the durability performance, the coating procedure may result in being useful with respect to both water absorption and alkaline aggression. Toledo et al. recommended the immersion of plant fibres in a silica fume slurry to improve such behaviour, before adding them into the mix [92]. The choice of coating treatment has to be accurately defined by taking into account not only durability aspects, but also the mechanical behaviour of the fibres themselves and with respect to the fibre-to-matrix bond. Fibre hornification treatments may also improve their durability by reducing the water absorption capacity [73].

The available literature data concerning the use of plant fibres in inorganic matrices shows a significant potential of such a reinforcement technique with respect to mechanical aspects. The experimental results, in some cases deriving by pioneering studies in the field, encourage to invest further efforts to improve the performance and to shed light on several aspects still unclear. Moreover, with respect to durability issues, it is important to spend more energy in improving the available treatments already existent, and eventually develop new ones. Finally, it is necessary to enrich the knowledge with respect to the efficiency of the application of Natural-TRMs as reinforcement of masonry, at scales of analysis being representative of the structural elements [78].

2.3 Aims of the Thesis

The state-of-the-art review emphasizes the importance in the development and improvement of innovative construction techniques, such as those characterised by the use of plant fibres-based materials, to provide sustainable solutions becoming more and more indispensable.

In the light of what it was presented in the previous sections, several outstanding issues still remain unsolved concerning the use of plant fibres as reinforcement in inorganic matrix composites. It is necessary, on the basis of the potential exhibited by such systems, and on the basis of their limitations with respect to conventional reinforcement systems, to intensify experimental analysis in order to have a more robust scientific background. However, such investigations have to be carefully conceived in a targeted manner with respect to the crucial aspects affecting either the strong points or the weaknesses of the reinforcing system.

In this context, the present thesis aims at providing a comprehensive investigation of a plant fibre-based TRM system. Specifically, due to its satisfying mechanical and physical properties, a flax textile was adopted as reinforcement of a hydraulic lime-based mortar. The overall interest of the work is to provide, by means of experimental investigations, a detailed analysis concerning the contribution of such composite systems on the mechanical behaviour of structural masonry elements.

However, too many aspects may play a more or less significant role in the global response of the system, and the existing studies in literature, although providing valuable information, are not enough to get a standardisation of the material. Therefore, in order to get a deep understanding of the overall mechanical behaviour, the investigation cannot be limited to the analysis of the response of masonry structural elements externally strengthened by Flax TRMs, but must start from more detailed scales of analysis.

The comprehensive characterisation carried out in the study moves from the physical investigation of the microstructure of the employed plant fibres, to the analysis of the global response of externally strengthened structural elements. For each scale of analysis, the main objectives pursued are defined in detail.

- At the smallest scale of analysis, a physical and morphological characterisation of the adopted plant fibres is presented aiming at providing the necessary information to evaluate the mechanical behaviour on greater scales of analysis, and to compare the adopted fibres with others of the same, or different, nature.
- A mechanical characterisation of the employed flax textile is proposed with the aim of defining to what extent it can be both mechanically and chemically suitable as reinforcement in inorganic matrices. Both mechanical and durability tests provide more information on such aspect and represent the basis from which possible necessary improvement solutions may be properly conceived.
- Mechanical and durability tests on the composite scale of analysis were planned with the aim of qualifying the material according to standardised procedures, and of comparing its performance with similar systems. At this scale of investigation

too, the analysis of results provides both potential aspects and limitations of the system.
- The global mechanical response of masonry assemblage aims at quantifying the contribution of the Flax TRM external strengthening system on the in-plane capacity of the panel. Such scale of analysis provides more realistic strength values with respect to smaller scales of investigation.

A further objective of the present study consists of taking advantage of the adopted multi-scale approach of the analysis to define the connection between the mechanical performance assessed on the material scale of investigation and the global response of the structural elements. This aspect is of fundamental importance since, according to standard codes concerning structural interventions by means of TRM systems, the main mechanical designing parameters are assessed on the composite scale of analysis. In addition, since only few studies in the literature have been carried out with a similar multi-level analysis approach, it represents a valuable contribution.

Therefore, in a global perspective, the comprehensive characterisation herein proposed has the aim of investigating the efficiency of a Flax TRM composite as external strengthening of masonry elements, by highlighting both the potentials and the limitations, with respect to a mechanical point of view.

Moreover, to move a step further than a mechanical characterisation for its own sake, the study aims to preliminary investigate technical solutions, i.e. fibre treatment and mortar mix design modification, to prove that the composite system, already characterised by a promising potential, still presents room for improvement.

References

1. AC 434–1011-R1 (2018) Acceptance criteria for masonry and concrete strengthening using fiber-reinforced cementitious matrix (FRCM) composite systems. ICC Evaluation Service, LLC: Birmingham, AL, USA
2. ACI 549.4R-13 (2013) Guide to design and construction of externally bonded fabric-reinforced cementitious matrix (FRCM) systems for repair and strengthening concrete and masonry structures; ACI Committee 549: Farmington Hills, MI, USA
3. Ambraseys N (2009) Earthquakes in the mediterranean and middle east: a multidisciplinary study of seismicity up to 1900. Cambridge University Press
4. Arboleda D, Carozzi FG, Nanni A, Poggi C (2016) Testing procedures for the uniaxial tensile characterization of fabric-reinforced cementitious matrix composites. J Compos Constr 20(3):04015063
5. Ascione L, de Felice G, De Santis S (2015) A qualification method for externally bonded Fibre Reinforced Cementitious Matrix (FRCM) strengthening systems. Compos B Eng 78:497–506
6. Askouni PD, Papanicolaou C (2018) Comparison of double-lap/double-prism and single-lap/single-prism shear tests for the TRM-to-Masonry bond assessment. Strain-Hardening Cement-Based Compos, Dordrecht, pp 527–534
7. ASTM E519–2: 2003 (2003) Standard test method for diagonal tension (shear) in masonry assemblages. ASTM committee C15 on manufactured Masonry Units, West Conshohocken, PA, US

8. Bilotta A, Ceroni F, Lignola GP, Prota A (2017) Use of DIC technique for investigating the behaviour of FRCM materials for strengthening masonry elements. Compos B Eng 129:251–270
9. Bisby L (2016) Fire resistance of textile fiber composites used in civil engineering. Textile fibre composites in civil engineering. Ed Woodhead Publishing, pp 169–185
10. Bisby L, Stratford T, Hart C, Farren S (2013) Fire performance of well-anchored TRM, FRCM and FRP flexural strengthening systems. Advanced composites in constructions, network group for composites in construction, Belfast, UK
11. Bisby L, Stratford T, Smith J, Halpin S (2011) FRP versus fiber reinforced cementitious mortar systems at elevated temperature. Spec Publ 275:1–20
12. Caggegi C, Carozzi FG, De Santis S, Fabbrocino F, Focacci F, Hojdys Ł, Lanoye E, Zuccarino L (2017a) Experimental analysis on tensile and bond properties of PBO and aramid fabric reinforced cementitious matrix for strengthening masonry structures. Compos B Eng 127:175–195
13. Caggegi C, Ferrara G, Lanoye E, Nguyen DB, Gabor A, Martinelli E (2017b) Experimental study on the effectiveness of masonry-basalt TRM reinforced systems characterized by different fiber grid densities. Key Eng Mater 747:266–273
14. Caggegi C, Lanoye E, Djama K, Bassil A, Gabor A (2017c) Tensile behaviour of a basalt TRM strengthening system: influence of mortar and reinforcing textile ratios. Compos B Eng 130:90–102
15. Carozzi FG, Poggi C (2015) Mechanical properties and debonding strength of Fabric Reinforced Cementitious Matrix (FRCM) systems for masonry strengthening. Compos B Eng 70:215–230
16. Carozzi FG, Bellini A, D'Antino T, de Felice G, Focacci F, Hojdys Ł, Laghi L, Lanoye E, Micelli F, Panizza M, Poggi C (2017) Experimental investigation of tensile and bond properties of Carbon-FRCM composites for strengthening masonry elements. Compos B Eng 128:100–119
17. Carozzi FG, Colombi P, Fava G, Poggi C (2016) A cohesive interface crack model for the matrix–textile debonding in FRCM composites. Compos Struct 143:230–241
18. Carus M, Scholz L (2011) Targets for bio-based composites and natural fibres. Biowerkstoff report, 8th ed, p 24. ISSN 1867–1217
19. Cevallos OA, Olivito RS (2015) Effects of fabric parameters on the tensile behaviour of sustainable cementitious composites. Compos B Eng 69:256–266
20. Cevallos OA, Olivito RS, Codispoti R, Ombres L (2015) Flax and polyparaphenylene benzobisoxazole cementitious composites for the strengthening of masonry elements subjected to eccentric loading. Compos B Eng 71:82–95
21. Charlet K, Eve S, Jernot JP, Gomina M, Breard J (2009) Tensile deformation of a flax fiber. Procedia Eng 1(1):233–236
22. CNR-DT 215/2018 (2018) Istruzioni per la Progettazione, l'Esecuzione ed il Controllo di Interventi di Consolidamento Statico mediante l'utilizzo di Compositi Fibrorinforzati a matrice inorganica, Consiglio Nazionale delle Ricerche—Commissione per lo studio per la predisposizione e l'analisi di norme tecniche relative alle costruzioni (Italy)
23. Codispoti R, Oliveira DV, Olivito RS, Lourenço PB, Fangueiro R (2015) Mechanical performance of natural fiber-reinforced composites for the strengthening of masonry. Compos B Eng 77:74–83
24. Coppola L et al (2018) Binders alternative to Portland cement and waste management for sustainable construction—part 2. J Biomater Appl 16(4):207–221
25. D'Almeida A, Filho JM, Filho RT (2009) Use of curaua fibers as reinforcement in cement composites. Chem Eng Trans 17:1717–1722
26. D'Ambrisi A, Feo L, Focacci F (2013) Experimental and analytical investigation on bond between Carbon-FRCM materials and masonry. Compos B Eng 46:15–20
27. D'Antino T, Papanicolaou C (2017) Mechanical characterization of textile reinforced inorganic-matrix composites. Compos B Eng 127:78–91

28. D'Antino T, Carloni C, Sneed LH, Pellegrino C (2014) Matrix–fiber bond behavior in PBO FRCM composites: a fracture mechanics' approach. Eng Fract Mech 117:94–111
29 de Silva AF, Butler M, Hempel S, Toledo Filho RD, Mechtcherine V (2014) Effects of elevated temperatures on the interface properties of carbon textile-reinforced concrete. Cement Concr Compos 48:26–34
30. de Carvalho Bello CB, Boem I, Cecchi A, Gattesco N, Oliveira DV (2019) Experimental tests for the characterization of sisal fiber reinforced cementitious matrix for strengthening masonry structures. Constr Build Mater 219:44–55
31. de Felice G, Aiello MA, Caggegi C, Ceroni F, De Santis S, Garbin E, Gattesco N, Hojdys Ł, Krajewski P, Kwiecień A, Leone M, Lignola GP, Mazzotti C, Oliveira D, Papanicolaou C, Poggi C, Triantafillou T, Valluzzi MR, Viskovic A (2018) Recommendation of RILEM technical committee 250-CSM: test method for textile reinforced mortar to substrate bond characterization. Mater Struct 51(4):95
32. de Gutiérrez RM, Díaz LN, Delvasto S (2005) Effect of pozzolans on the performance of fiber-reinforced mortars. Cement Concr Compos 5(27):593–598
33. De Santis S, de Felice G (2014) Tensile behavior and durability of mortar-based strengthening systems with glass-aramid textiles. Key Eng Mater 624:346–353
34. De Santis S, de Felice G (2015) Tensile behaviour of mortar-based composites for externally bonded reinforcement systems. Compos B Eng 68:401–413
35. De Santis S, Carozzi FG, de Felice G, Poggi C (2017a) Test methods for textile reinforced mortar systems. Compos B Eng 127:121–132
36. De Santis S, Ceroni F, de Felice G, Fagone M, Ghiassi B, Kwiecień A, Lignola GP, Morganti M, Santandrea M, Valluzzi MR, Viskovic A (2017b) Round Robin test on tensile and bond behaviour of steel reinforced grout systems. Compos B Eng 127:100–120
37. Del Zoppo M, Di Ludovico M, Prota A (2019) Analysis of FRCM and CRM parameters for the in-plane shear strengthening of different URM types. Compos B Eng 171:20–33
38. Dicker MPM, Duckworth PF, Baker AB, Francois G, Hazzard MK, Weaver PM (2014) Green composites: a review of material attributes and complementary applications. Compos a Appl Sci Manuf 56:280–289
39. Dittenber DB, GangaRao HVS (2012) Critical review of recent publications on use of natural composites in infrastructure. Compos Appl Sci Manuf 43(8):1419–1429
40. Donnini J (2019) Durability of glass FRCM systems: effects of different environments on mechanical properties. Compos B Eng 174:107047
41. Donnini J, Caso De, y Basalo F, Corinaldesi V, Lancioni G, Nanni A (2017) Fabric-reinforced cementitious matrix behavior at high-temperature: experimental and numerical results. Compos B Eng 108:108–121
42. Donnini V, Corinaldesi V (2017) Mechanical characterization of different FRCM systems for structural reinforcement. Constr Build Mater 145:565–575
43. Ferrara G, Martinelli E (2018) Tensile behaviour of textile reinforced mortar composite systems with flax fibres. In: Proceedings of the 12th fib international PhD symposium in civil engineering, pp 863–869
44. Ferrara G, Caggegi C, Gabor A, Martinelli E (2019a) Experimental study on the adhesion of basalt textile reinforced mortars (TRM) to clay brick masonry: the influence of textile density. Fibers 7(12):103
45. Ferrara G, Caggegi C, Gabor A, Martinelli E (2019b) Mechanical behaviour of masonry panels strengthened by flax TRM systems. Key Eng Mater 817:427–434
46. Ferrara G, Caggegi C, Martinelli E, Gabor A (2020) Shear capacity of masonry walls externally strengthened using Flax-TRM composite systems: experimental tests and comparative assessment. Constr Build Mater 261:120490
47. Ferreira SR, de Silva AF, Lima PRL, Toledo Filho RD (2015) Effect of fiber treatments on the sisal fiber properties and fiber–matrix bond in cement based systems. Constr Build Mater 101:730–740
48. Ferreira SR, Martinelli E, Pepe M, de Andrade Silva F, Toledo Filho RD (2016) Inverse identification of the bond behavior for jute fibers in cementitious matrix. Compos B Eng 95:440–452

49. Fidelis MEA, Toledo Filho RD, de Andrade Silva F, Mobasher B, Müller S, Mechtcherine V (2019) Interface characteristics of jute fiber systems in a cementitious matrix. Cem Concr Res 116:252–265
50. Franzoni E, Gentilini C, Santandrea M, Zanotto S, Carloni C (2017) Durability of steel FRCM-masonry joints: effect of water and salt crystallization. Mater Struct 50(4):201
51. Geissdoerfer M, Savaget P, Bocken NMP, Hultink EJ (2017) The circular economy—a new sustainability paradigm? J Clean Prod 143:757–768
52. Ghiassi B, Razavizadeh A, Oliveira DV, Marques V, Lourenço PB (2015) Tensile and bond characterization of natural fibers embedded in inorganic matrices. In: Proceeding of 2nd international conference on natural fibers, Azores/Portugal, 27–29 Apr 2015
53. Giacomin G (2016) Innovative strengthening materials for the post-earthquake reconstruction of L'Aquila masonries. In: Proceedings of the 10th international conference on structural analysis of historical constructions, SAHC 2016, LEUVEN, Belgium
54. Gram HE (1983) Durability of natural fibres in concrete. Swedish cement and concrete research institute of technology, Stockholm, Sweden
55. Harajli M, ElKhatib H, San-Jose JT (2010) Static and Cyclic out-of-plane response of masonry walls strengthened using textile-mortar system. J Mater Civ Eng 22(11):1171–1180
56. Indirli M, Kouris LAS, Formisano A, Borg RP, Mazzolani FM (2013) Seismic damage assessment of unreinforced masonry structures after The Abruzzo 2009 earthquake: the case study of the historical centers of L'Aquila and Castelvecchio Subequo. Int J Arch Herit 7(5):536–578
57. Kaplan H, Bilgin H, Yilmaz S, Binici H, Öztas A (2010) Structural damages of L'Aquila (Italy) earthquake. Nat Hazards Earth Syst Sci 10(3):499–507
58. Komuraiah A, Kumar NS, Prasad BD (2014) Chemical composition of natural fibers and its influence on their mechanical properties. Mech Compos Mater 50(3):359–376
59. Kouris LAS, Triantafillou TC (2018) State-of-the-art on strengthening of masonry structures with textile reinforced mortar (TRM). Constr Build Mater 188:1221–1233
60. Leone M, Aiello MA, Balsamo A, Carozzi FG, Ceroni F, Corradi M, Gams M, Garbin E, Gattesco N, Krajewski P, Mazzotti C, Oliveira D, Papanicolaou C, Ranocchiai G, Roscini F, Saenger D (2017) Glass fabric reinforced cementitious matrix: tensile properties and bond performance on masonry substrate. Compos B Eng 127:196–214
61. Lignola GP, Caggegi C, Ceroni F, De Santis S, Krajewski P, Lourenço PB, Morganti M, Papanicolaou C, Pellegrino C, Prota A, Zuccarino L (2017) Performance assessment of basalt FRCM for retrofit applications on masonry. Compos B Eng 128:1–18
62. Linea Guida per la identificazione (2018) la qualificazione ed il controllo di accettazione di compositi fibrorinforzati a matrice inorganica (FRCM) da utilizzarsi per il consolidamento strutturale di costruzioni esistenti. Consiglio Superiore Lavori Pubblici (Italy)
63. Liu Q, Stuart T, Hughes M, Sharma HSS, Lyons G (2007) Structural biocomposites from flax—Part II: The use of PEG and PVA as interfacial compatibilising agents. Compos a Appl Sci Manuf 38(5):1403–1413
64. Menna C, Asprone D, Durante M, Zinno A, Balsamo A, Prota A (2015) Structural behaviour of masonry panels strengthened with an innovative hemp fibre composite grid. Constr Build Mater 100:111–121
65. Mercedes L, Gil L, Bernat-Maso E (2018) Mechanical performance of vegetal fabric reinforced cementitious matrix (FRCM) composites. Constr Build Mater 175:161–173
66. Meyer C (2009) The greening of the concrete industry. Cement Concr Compos 31(8):601–605
67. Mezrea PE, Yilmaz IA, Ispir M, Binbir E, Bal IE, Ilki A (2017) External Jacketing of Unreinforced Historical Masonry Piers with Open-Grid Basalt-Reinforced Mortar. J Compos Construct 21(3):04016110
68. Micelli F, Aiello MA (2019) Residual tensile strength of dry and impregnated reinforcement fibres after exposure to alkaline environments. Compos B Eng 159:490–501
69. Michels J, Motavalli M (2013) Strength evolution after high temperature exposure of coated carbon fiber rovings used for a fiber-reinforced cementitious matrix (FRCM). Fourth Asia pacific conference on FRP in structures (APFIS 2013), Melbourne, Australia

70. Mohanty AK, Misra M, Drzal LT, Selke SE, Harte BR, Hinrichsen G (2005) Natural fibres. In: Boca Raton FL (ed) biopolymers, and biocomposites: an introduction in natural fibres, natural fibres, biopolymers, and biocomposites. CRC Press, Taylor & Francis Group, pp 1–36
71. Motta LA, John VM, Agopyan V (2010) Thermo-mechanical treatment to improve properties of sisal fibres for composites. Mater Sci Forum 636–637:253–259
72. Nobili A (2016) Durability assessment of impregnated Glass Fabric Reinforced Cementitious Matrix (GFRCM) composites in the alkaline and saline environments. Constr Build Mater 105:465–471
73. Nobili A, Signorini C (2017) On the effect of curing time and environmental exposure on impregnated Carbon Fabric Reinforced Cementitious Matrix (CFRCM) composite with design considerations. Compos B Eng 112:300–313
74. NTC 2018 (2018) Norme Tecniche per le Costruzioni, Ministry of the infrastructures and transports of the Italian Republic.
75. Olivito RS, Cevallos OA, Carrozzini A (2014) Development of durable cementitious composites using sisal and flax fabrics for reinforcement of masonry structures. Mater Des 57:258–268
76. Olivito RS, Codispoti R, Cevallos OA (2016) Bond behavior of Flax-FRCM and PBO-FRCM composites applied on clay bricks: experimental and theoretical study. Compos Struct 146:221–231
77. Olivito RS, Dubois F, Venneri A, Zuccarello FA (2012) Experimental and numerical analysis of masonry macroelements reinforced by natural-fibre-composite materials. In: Proceedings of 4th international conference on FRP composites in civil engineering, rome/italy, 13–15 Jun 2012
78. Pacheco-Torgal F, Jalali S (2011) Cementitious building materials reinforced with vegetable fibres: a review. Constr Build Mater 25(2):575–581
79. Papanicolaou C, Triantafillou T, Lekka M (2011) Externally bonded grids as strengthening and seismic retrofitting materials of masonry panels. Constr Build Mater 2(25):504–514
80. Papanicolaou CG, Triantafillou TC, Karlos K, Papathanasiou M (2007) Textile-reinforced mortar (TRM) versus FRP as strengthening material of URM walls: in-plane cyclic loading. Mater Struct 40(10):1081–1097
81. Papanicolaou CG, Triantafillou TC, Papathanasiou M, Karlos K (2008) Textile reinforced mortar (TRM) versus FRP as strengthening material of URM walls: out-of-plane cyclic loading. Mater Struct 41(1):143–157
82. Ramakrishna G, Sundararajan T (2005) Studies on the durability of natural fibres and the effect of corroded fibres on the strength of mortar. Cement Concr Compos 27(5):575–582
83. Ramamurthi DS, Sophia M (2016) A review on modified lime-based mortars—an alternative to cement mortar. Int J Innov Sci Eng Technol 2(12):350–356
84. RILEM TC IMC: Durability of inorganic matrix composites used for strengthening of masonry constructions. https://www.rilem.net/groupe/imc-durability-of-inorganic-matrix-composites-used-for-strengthening-of-masonry-constructions-384. Accessed 7th Oct 2020
85. Technical Committee RILEM, 232-TDT (Wolfgang Brameshuber) (2016) Recommendation of RILEM TC 232-TDT: test methods and design of textile reinforced concrete. Mater Struct 49(12):4923–4927
86. Rizzo V, Bonati A, Micelli F, Leone M, Aiello MA (2019) Influence of Alkaline environments on the mechanical properties of FRCM/CRM and their materials. Key Eng Mater 817:195–201
87. Saidi M, Gabor A (2019) Use of distributed optical fibre as a strain sensor in textile reinforced cementitious matrix composites. Measurement 140:323–333
88. Saidi M, Gabor A (2020) Experimental analysis of the tensile behaviour of textile reinforced cementitious matrix composites using distributed fibre optic sensing (DFOS) technology. Constr Build Mater 230:117027
89. Signorini C, Nobili A, Cedillo González EI, Siligardi C (2018) Silica coating for interphase bond enhancement of carbon and AR-glass Textile Reinforced Mortar (TRM). Compos B Eng 141:191–202

90. Sneed LH, D'Antino T, Carloni C, Pellegrino C (2015) A comparison of the bond behavior of PBO-FRCM composites determined by double-lap and single-lap shear tests. Cement Concr Compos 64:37–48
91. Tetta ZC, Bournas DA (2016) TRM versus FRP jacketing in shear strengthening of concrete members subjected to high temperatures. Compos B Eng 106:190–205
92. Tolêdo Filho RD, Ghavami K, England GL, Scrivener K (2003) Development of vegetable fibre–mortar composites of improved durability. Cement Concr Compos 25(2):185–196
93. Tolêdo Filho RD, Scrivener K, England GL, Ghavami K (2000) Durability of alkali-sensitive sisal and coconut fibres in cement mortar composites. Cement Concr Compos 22:127–143
94. Tonoli GHD, Joaquim AP, Arsène MA, Bilba K Jr, Savastano H (2007) Performance and durability of cement based composites reinforced with refined Sisal Pulp. Mater Manuf Process 22(2):149–156
95. Tran MT, Vu XH, Ferrier E (2018) Experimental and analytical analysis of the effect of fibre treatment on the thermomechanical behaviour of continuous carbon textile subjected to simultaneous elevated temperature and uniaxial tensile loadings. Constr Build Mater 183:32–45
96. Trapko T (2013) The effect of high temperature on the performance of CFRP and FRCM confined concrete elements. Compos B Eng 54:138–145
97. Triantafillou T (2011) Innovative textile-based composites for strengthening and seismic retrofitting of concrete and masonry structures. Advances in FRP composites in civil engineering, Berlin, Heidelberg, pp 3–12
98. Tumialan JG, Micelli F, Nanni A (2001) Strengthening of masonry structures with FRP composites. Struct Cong 1–8
99. UN (1987) Report of the world commission on environment and development: our common future, UN documents, transmitted to the general assembly as an annex to document A/42/427—development and international Co-operation: environment. https://www.un-documents.net/wced-ocf.htm. Accessed 8th Oct 2020
100. Van Den Einde L, Zhao L, Seible F (2003) Use of FRP composites in civil structural applications. Constr Build Mater 17(6):389–403
101. Williams PN, Flansbjer M, Johannesson P, Malaga K, Lundgren K (2016) Tensile behaviour of textile reinforcement under accelerated ageing conditions. J Build Eng 5:57–66
102. Yan L, Chouw N, Jayaraman K (2014) Flax fibre and its composites—a review. Compos B Eng 56:296–317
103. Yan L, Kasal B, Huang L (2016) A review of recent research on the use of cellulosic fibres, their fibre fabric reinforced cementitious, geo-polymer and polymer composites in civil engineering. Compos B Eng 92:94–132
104. Zhuge Y, Chuang SW (2005) Seismic retrofitting of unreinforced masonry buildings—a literature review. Aust J Struct Eng 6:25–36

Chapter 3
Mechanical Behaviour and Durability of Flax Textiles to Be Used as Reinforcement in TRMs

Prior to addressing the investigation of Natural TRMs, and to studying its influence on structural elements, it is necessary to characterise the reinforcing fabric. Such step, conducted at the constituent "*materials*" scale of analysis, is of fundamental importance for the interpretation of the mechanical behaviour at greater scales of analysis. Due to the randomness of plant fibre properties, it is not enough to limit such investigation to the assessment of mechanical properties, such as tensile strength and stiffness. It is, however, essential to accurately define the geometry and the morphology of the fibres, and to face the issue of the compatibility of the textile with the mortar. The aim of this chapter is to provide such information by reporting the results of an experimental research intended at determining the fundamental properties of Flax fibres, yarns and fabrics, and assess their behaviour when subjected to various environmental exposures and ageing protocols. The experimental work moves from determining the relevant geometric properties, such as length, cross-section and relevant sizes, and physical parameters, such as water absorption capacity. Then, tensile tests on differently sized flax samples lead to determining the tensile strength and its natural variability. Lastly, regarding durability, different ageing protocols are considered with the aim of reproducing specific environmental conditions, i.e. the exposure to water, salt-water, and alkali solutions or to lime- and cement-based mortars.

The experimental activity performed to physically and mechanically characterise the flax textile, and to investigate its durability were almost entirely carried out at the Department of Industrial Engineering (DIIN) of University of Salerno. Few experimental investigations were performed at the laboratory NUMATS-COPPE of the Federal University of Rio de Janeiro in order to deeply analyse such aspects.

G. Ferrara, *Flax-TRM Composite Systems for Strengthening of Masonry*, Springer Theses,
https://doi.org/10.1007/978-3-030-70273-1_3

3.1 Physical Characterisation

The flax textile adopted in the research was supplied by the company Innovations s.r.l. and named as *FIDFLAX Grid 300 HS20®* [14]. It is a bi-directional woven textile characterised by plain wave. Yarns are arranged in warp and weft directions by forming a simple cross pattern. Yarns are laid in twos along the same row in both the directions. Each one is characterised by two smaller threads intertwined in a spiral-like pattern. Yarns are in their turn a combination of filaments, considered as the minor constituent of the fabric. In this study, the intertwined thread represents the reference sample, hereinafter it is referred to simply as thread. The structure of the flax textile at different scale levels is represented in Fig. 3.1.

The fabric is characterised by 4.3 threads per centimetre in both directions, and a clear opening size of about 3. This size was accurately defined to use the textile as reinforcement within inorganic matrices, allowing the transit of fine aggregates through the clear spaces [5]. The identification of physical and geometric properties starts from analysing the single filaments constituting the fabric yarns, up to the assessment of the thread's density and cross section. The size of the filaments is a significant parameter to identify plant fibres physically and mechanically. To assess the filament diameter images obtained by a Scanning Electron Microscope (SEM) analysis were utilised (Fig. 3.2). A number of 4 SEM images representing 4 different samples, and with a magnification of 1000X, were processed providing the filament diameter as the mean of 28 measurements.

The estimation of the thread cross-section was indirectly obtained from the value of the density of the textile. A number of 5 saturated flax threads, with a length of 150 mm, were considered to estimate the specific gravity by means of a hydrostatic balance. The thread mean cross-section was estimated by dividing the weight of a sample for its density and length.

The water absorption rate of the flax threads was estimated as well (Fig. 3.3). A number of 5 samples, having a length of 150 mm, and immersed in deionised water, were weighted with regular intervals of 24 h until a negligible weight gain was attained. In line with similar studies from literature [9], the eater absorption ratio was equal to 222%, with a coefficient of variation of 19%.

Table 3.1 lists the main values of flax textile physical properties with the respective coefficients of variation.

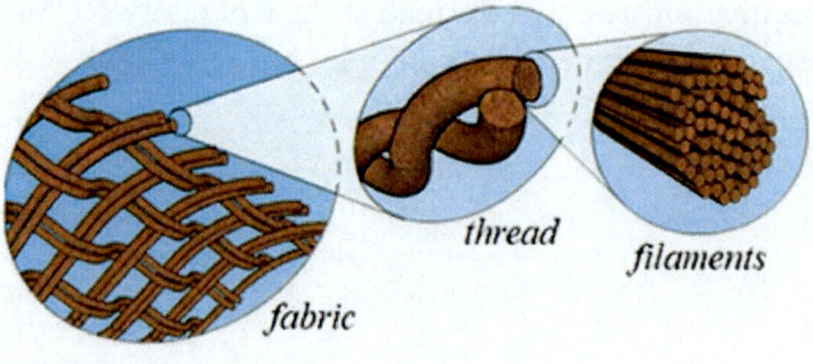

Fig. 3.1 Flax fabric structure

Fig. 3.2 **a** Flax fabric; **b** thread (magnification 100X); **c** filaments (magnification 1000X)

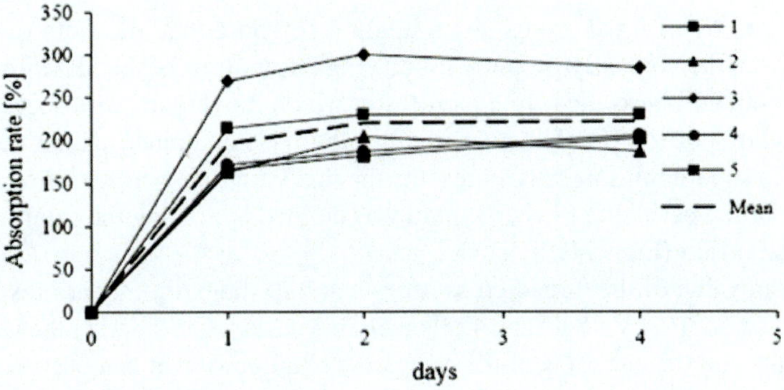

Fig. 3.3 Flax thread water absorption rate

Table 3.1 Flax fabric physical properties

	Mean	Co.V. (%)
Filament diameter (μm)	16.78	29.64
Density (g/cm^3)	1.19	3.29
Linear density (Tex)	302	15.27
N° threads/cm	4.3	–
Thread's cross-section area (mm^2)	0.25	16.62

A more detailed investigation of the morphology of the flax threads was carried out at the laboratory NUMATS-COPPE of the Federal University of Rio de Janeiro. The transversal shape of the threads was analysed by processing SEM images having a magnification of 60X. A total of 30 images were processed. The images are characterised by the effective transversal section of the threads, and by a number of voids, due to the chaotic arrangement of the filaments within the yarns, that prevent an immediate accurate estimation of the cross-section. For this reason, the SEM pictures

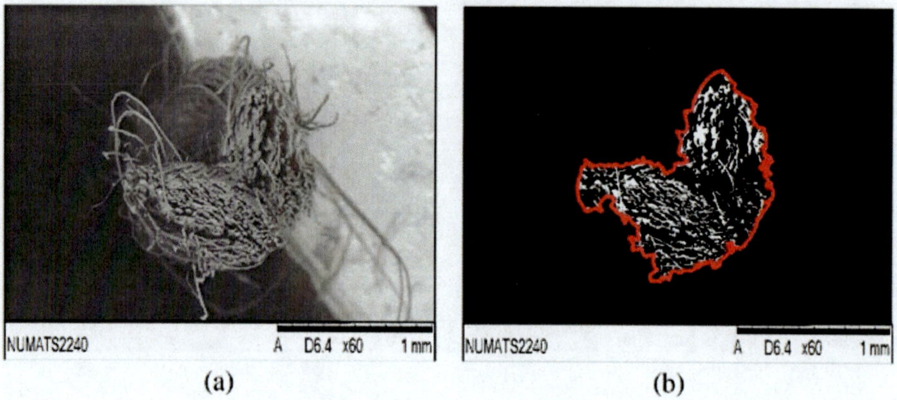

Fig. 3.4 Thread's morphology identification: **a** SEM image; **b** manipulated image

were analysed by using an image manipulation software that allows to distinguish the fibres from the voids. By properly modifying the contrast of the SEM images it is possible to obtain black-and-white pictures in which the thread cross-section represents the white portion [5]. The number of with pixels contained within the image, converted in square millimetres, represents the flax yarn transversal section area. An estimation of the perimeter of each thread was defined by tracing the external border of the with portion (Fig. 3.4).

The mean value of the thread cross section and of the perimeter are respectively equal to 0.28 mm^2 (Co.V. 18%) and 4.89 mm (Co.V. 21%). Considering the variability of the results and the accuracy of the adopted techniques it can be observed that the cross section mean value estimated by means of analysis on SEM images is in line with the indirect estimation obtained from the density of the fibres.

3.2 Mechanical Characterisation

3.2.1 Methods

The mechanical characterisation of flax textile mainly consists in the assessment of its response in tension. Plant fibre fabrics typically present a significant variability in the mechanical properties that also depends on the arrangement of the woven threads. This is the reason why the characteristic properties of the specific fibres utilised are generally assessed on the single element constituting the thread. With the aim to compare the performance of the adopted flax fibres with those of other types of plant fibres from literature, tensile tests were preliminary carried out on the single filaments, representing the smallest component of the flax textile. Moreover, the single flax tows were tested in tension to get mechanical properties more representative of the scale of analysis of the composite itself. Since the size of the sample may itself affect the

Fig. 3.5 Flax filament tested in tension

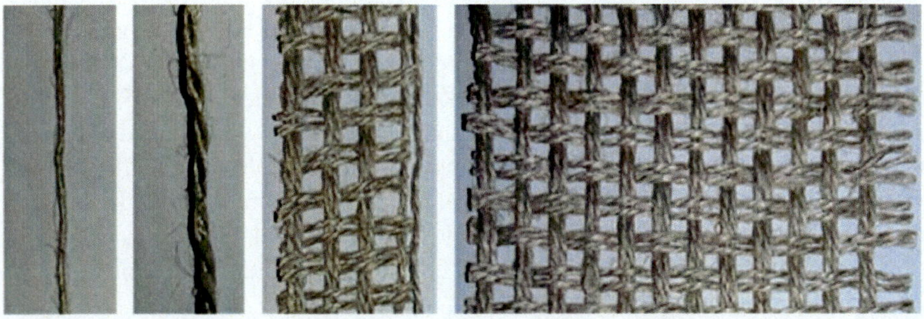

(a) (b) (c) (d)

Fig. 3.6 Flax samples: **a** Flax-Y; **b** Flax-T; **c** Flax fabric-2 cm; **d** Flax fabric-6 cm

estimation, different sized samples were considered to estimate the tensile strength and to analyse the variability of the mechanical proprieties with the size. The 5 different series considered, differing in the number of yarns, are defined hereafter:

- *Filament*: it consists of the smaller component of the fabric. The samples, 30 mm long, were randomly extracted from the fabric (Fig. 3.5);
- *Flax-Y*: it represents one of the two yarns constituting the main threads. The samples, 150 mm long, were extracted from the fabric in the weft direction (Fig. 3.6a);
- *Flax-T (warp and weft)*: they represent the thread defined as the main element of the fabric. The samples, 150 mm long, were extracted from the fabric in both warp and weft directions (Fig. 3.6b);
- *Flax fabric-2 cm*: it represents a flax fabric strip containing 8 main threads. The samples, 20 mm large and 150 mm long, were extracted from the fabric in the weft direction which (Fig. 3.6c). This configuration was chosen on the basis of having a specimen comprising at least 3 yarns in the strip;
- *Flax fabric-6 cm (warp and weft)*: they represent flax fabric strips containing 24 main threads. The samples, 60 mm large and 300 mm long, were extracted from the fabric in both warp and weft directions (Fig. 3.6d).

The filament series consisted of 10 samples with a gauge length of 10 mm that were clamped at each edge for a length of 10 mm. The tests were carried out at the laboratory NUMATS-COPPE of the Federal University of Rio de Janeiro by a microforce testing machine (Tytron 250, MTS System Corporation, Eden Prairie,

MN, USA), with 50 N load cell in displacement control by using a rate of 0.5 mm/min. The samples of the series Flax-T and Flax fabric-6 cm were extracted from both the textile directions to investigate any direction-related variability. The samples of the series Flax-Y, Flax-T and Flax fabric-2 cm had a gauge length of 100 mm and were clamped for a length of 25 mm along each edge. The test machine was provided with rough clamping elements to guarantee a proper gripping of the sample. Tensile tests were carried out in displacement control with a rate of 4 mm/min, using a 1 kN load cell, by means of a CMT4000 SANS Series dynamometer machine (by MTS, China), at the Department of Industrial Engineering of University of Salerno (Fig. 3.7a).

The samples of the series Flax fabric-6 cm had a gauge length of 200 mm and were clamped for a length og 50 mm along each edge. Two steel plates, 5 mm thick, between which the textile was glued by means of epoxy resin, were placed within the machine clamps guaranteeing a proper gripping of the samples (Fig. 3.7b). Tensile tests were carried out in displacement control with a rate of 4 mm/min, and according to the ISO 13934-1 [7]; by means of a Zwick Roell Schenck Hydropuls S56, with a maximum capacity of 630 kN, at the STRENGTH laboratory of University of Salerno.

The mean values of the axial strains were assessed by dividing the displacement between the clamping edges for the free length of the specimens.

(a) (b)

Fig. 3.7 Tensile test set-up of series: **a** Flax-T and **b** Flax fabric-6 cm

3.2.2 Results and Discussion

The displacement in the load direction and the value of the force were recorded during the tests. The results are reported in terms of stress and strain, easily obtained from a conversion of the recorded data. The parameters chosen to characterise the tensile behaviour of the fibres were the tensile strength, f_t, the Young's modulus, E, and the strain at the peak, ε_u. The stress-strain tensile curve is always characterised by a lower initial stiffness that increases with the strain up to a constant value that is kept until the rupture is attained. This behaviour is due to morphologic (microstructure of the fibres) and geometric (fabric arrangement) reasons. Within the study the Young's modulus is defined as the stress-strain curve slope in the linear branch comprised between the stress range from 20 to 50% of the maximum strength.

The stress-strain tensile response of the filaments is shown in Fig. 3.8. The cross-section area considered to get the stress values of filaments was obtained by assuming a circular section and by considering the diameter value reported in Table 3.1.

The stress-strain response of the other series of specimens is shown in Fig. 3.9. The stress value was obtained by dividing the applied load for the transversal area. The cross section was estimated on the basis of the value obtained from the threads of the Flax-T series, considering that a thread comprises 2 yarns, and that Flax-fabric-2 cm and Flax-fabric-6 cm respectively comprise 8 and 24 threads. Table 3.2 lists the main values of the parameters describing the tensile behaviour, with their respective coefficient of variation.

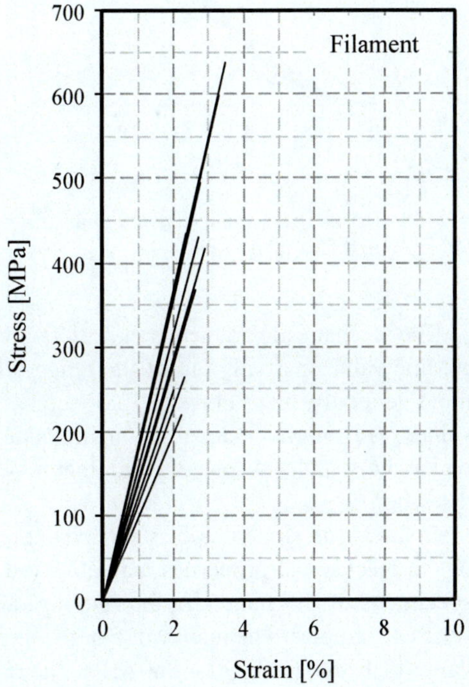

Fig. 3.8 Tensile stress-strain diagram of the flax filament series

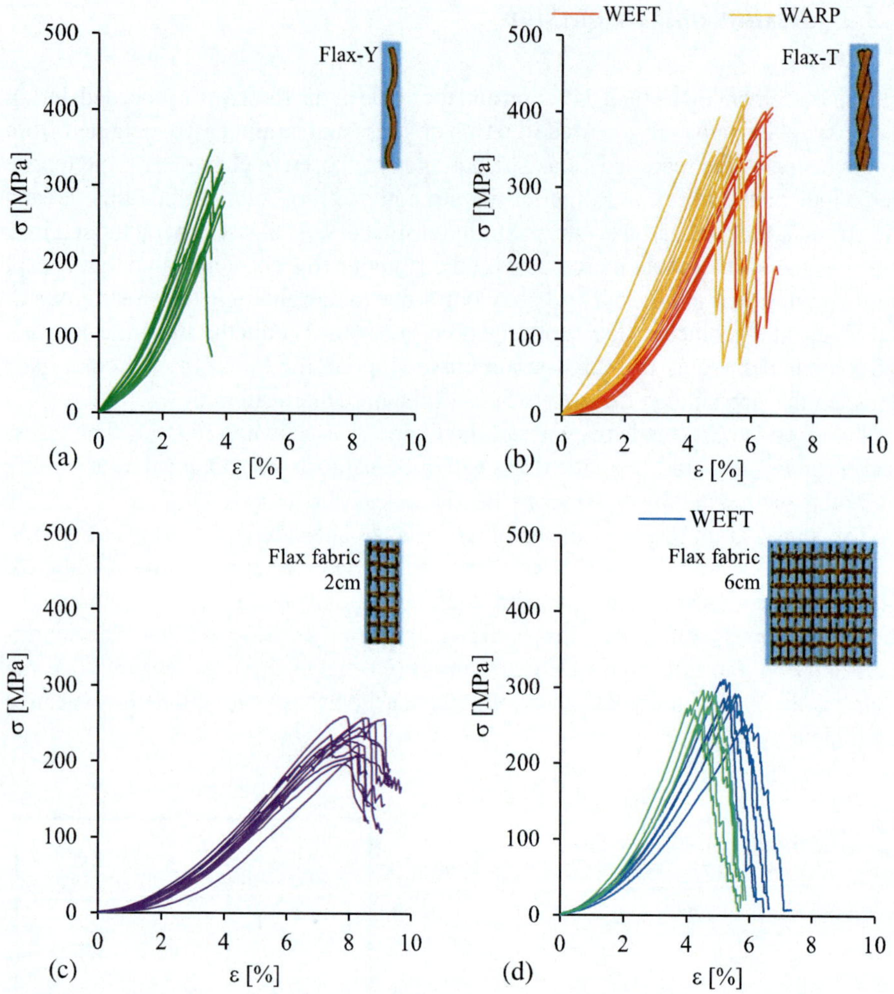

Fig. 3.9 Tensile stress-strain response of the different flax specimens: **a** Flax-Y; **b** Flax-T; **c** Flax fabric 2 cm; **d** Flax fabric 6 cm

The mechanical parameters that arose from the experimental tests on the filaments assume values partially out of the range of values found in literature for flax, and more generally, plant fibres [16]. Specifically, the Young's modulus presents lower values. This might be due to the manufacturing process of the filament roving from which the samples were extracted, that may have made the fibres to be subjected to distortion.

In lines with similar studies carried out on different plant fibres [12], a variability of the mechanical properties was observed. This variability occurred both among specimens of the same size and among the main values concerning the different series of samples. The main value of the tensile strength concerning the flax threads varies in the range of $290 \div 360$ MPa. The strength concerning the fabric strips varies

Table 3.2 Mechanical properties of flax samples tested in tension

Direction		Numb. of spec.	P_u		f_t		E		ε_u	
			Mean (N)	Co.V. (%)	Mean (MP)	Co.V. (%)	Mean (GPa)	Co.V. (%)	Mean (%)	Co.V. (%)
Filament	–	10	0.29	35	411	35	15.1	21	2.66	18
Flax-Y	Weft	15	37	16	293	16	11.61	10	3.58	12
Flax-T	Weft	12	88	13	345	13	8.53	6	6.14	9
Flax-T	Warp	14	92	10	361	10	10.07	7	5.20	11
Fabric-2 cm	Weft	14	476	9	233	9	4.39	8	8.27	6
Fabric-6 cm	Weft	5	1759	7	288	7	8.25	10	5.56	6
Fabric-6 cm	Warp	5	1744	4	285	4	9.81	3	4.63	5

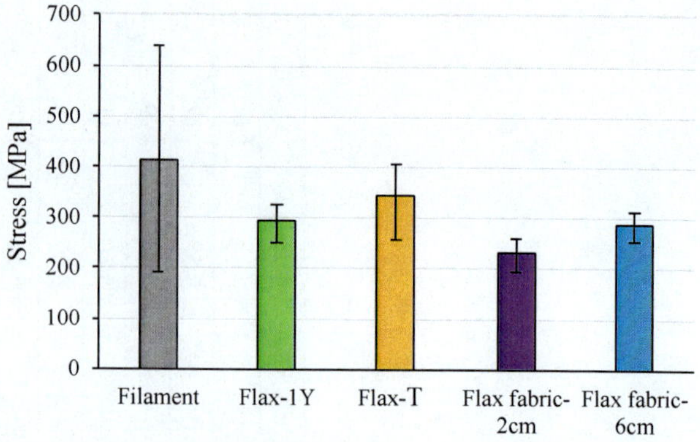

Fig. 3.10 Tensile strength of different sized falx samples

in the range of 230 ÷ 290 MPa. In both types of specimens, the Young's modulus attained values in the range of 8.3 ÷ 11.6 GPa, except for the specimens of the series Flax-fabric-2 cm that exhibited a much lower stiffness. Such inconsistency may be attributed to an inefficient clamping system adopted during the test. For this reason, further analysis would be needed to get more certain values of stiffness and strain parameters concerning the Flax-fabric-2 cm series.

Both strength and stiffness resulted to be lower with respect to the values deriving from the tests on the flax filaments. Such trend is in line with research carried out by using flax textile in which a loss of stiffness is recorded as well by increasing the size of the samples [3]. The strain at failure was comprised in the range 3.6 ÷ 6.1. The strain values are in line with studies from literature, specifically in the case of experimental analysis carried out on plant fibres thread, rather than on single filaments.

In Fig. 3.10 the mean shows the main values of the tensile strength concerning the different series investigated. The highest value of the tensile strength, equal to 411 MPa, is attained by the filaments, while the samples of the Flax-T series assume a mean value of 354 MPa. The latter, being assessed on threads meant to reinforce matrices, can be considered as the nominal strength of the textile adopted as reinforcement in inorganic matrix composites. The specimens having a larger size, notably concerning the Flax-fabric-2 cm and Flax-fabric-6 cm series, exhibited lower values of strength due to a non-uniform distribution of the stresses among the different threads, yarns and filaments comprised within the strip. However, the coefficient of variation concerning the tensile strength significantly decreases by increasing the size of the specimens emphasising a lower variability of the response in the larger samples (Table 3.2). The mechanical behaviour of the series Flax fabric-6 cm is the most representative of the textile adopted in TRM applications (and for instance in Flax TRM tensile tests). The tensile response of the strip is characterised by an initial lower stiffness up to a quasi-linear branch in which the stress-strain curve slope assumes a constant value (Fig. 3.9d). The less stiff initial branch is due, as discussed in the literature review section, to the microstructure of the filaments, and to the arrangement of the filaments within the yarns. However, a third contribution is due to the structure of the fabric itself, characterised by yarns differently loaded due to the non-uniform distribution of the stresses. The mean axial strain corresponding to the failure is equal to 5%. It is important to underline that this value take into account also of the initial less stiff behaviour of the textile. In the applications, such behaviour can be modified in order to have a linear elastic response of the textile.

The experimental analysis showed above provides the main mechanical parameters concerning the flax textile. Such values are of fundamental importance in designing process of composite materials reinforced with flax fabric.

3.3 Durability of Flax Fibres

The falx fabric adopted in the study is meant to be used as reinforcement in mortar based (lime or cement based) composites. In case of strengthening of historic masonry structures, the materials adopted for the intervention must be consistent with those constituting the reinforced building. In these cases, hydraulic lime-based composites are preferred, while in case of no specific restriction related to the nature of the structure the use of cement mortars may be more appropriate.

When applied in mortar-based environments, plant fibres give rise to durability problems affecting their mechanical properties in the long term. Notably, the alkaline environment arises two main ageing effects: the alkaline hydrolysis and the mineralisation process of fibre constituents [15]. To better understand these degrading processes, it is necessary to make attention on the fibre structure. As known, the plant fibre main constituents are:

- cellulose, it represents the most important structural component;
- hemicellulose, typically present along with cellulose;
- lignin, representing the binding phase for cellulose.

While cellulose has a crystalline and strong structure, hemicellulose is characterised by random and amorphous structures with lower strength. Mortars comprise a high alkaline water content that causes the dissolution of hemicellulose and lignin reducing the degree of polymerisation of the fibre, with the consequent alkaline hydrolysis of cellulose cells. The mineralisation process, instead, consists in the migration of hydration products, such as $Ca(OH)_2$, into the lumen and voids of the fibre structure [6].

3.3.1 Methods

This section aims at assessing the long-term mechanical behaviour of plant fibres adopted in mortar-based environments. To this purpose, three ageing protocols are considered. Flax threads were tested after a specific period of exposure in an ageing environment. After the removal from the exposure environment the samples were dried in environment conditions for 24 h before being tested. The samples considered in this study consists in flax threads (defined before as Flax-T), 150 mm long, tested in tension in displacement control with a rate of 4 mm/min, via a SANS Universal Testing Machine with 10 kN maximum capacity. The samples had a gauge length of 100 mm. A number of samples between 12 and 15 was tested for each series. Flax-T samples showed in the previous section were considered to define the reference behaviour of untreated flax fibres. The degrade state of the fibres after the exposure periods was investigated by the analysis of SEM images. The implementation of ageing protocols, the tests and the SEM analysis were executed at the Department of Industrial Engineering of University of Salerno.

3.3.1.1 Ageing Protocol 1

Acceptance criteria concerning fabric reinforced cementitious matrix composite systems [1], define durability requirements the material has to comply with. Notably, it defines the ageing environments, the exposure period length, and target limits in terms of strength decay.

Three different ageing environments are described: water, saltwater (reproducing marine environment), and alkaline water. As explained before, this section deals with the durability of plant fibres to be employed in composites. As preliminary step of a more comprehensive study to be carried out on the entire composite, the fabric is herein subjected to the ageing protocols conceived for the composite. The length of the exposure period was fixed at 1000 h. The ageing protocol comprises the environment conditions described as follow:

- *Flax-T Water 1000*: the threads were immersed in 500 ml of deionised water for a duration of 1000 h (~42 days);
- *Flax-T Salt 1000*: the threads were immersed in 500 ml of saline solution of deionised water with a concentration of 3.5% in weight of NaCl for a duration of 1000 h (~42 days);
- *Flax-T Alk 1000*: the threads were immersed in 500 ml of alkaline solution with a pH of 9.5 obtained by adding to deionised water NaOH (concentration of 0.32% in weight), for a period of 1000 h (~42 days).

The ageing was carried out at laboratory temperature, 20 °C, and relative humidity, 50%.

3.3.1.2 Ageing Protocol 2

The ageing protocol 2 represents a preliminary study aiming at assessing the decay of mechanical properties of flax threads once embedded in mortars. On the basis of similar studies in the literature [15], flax threads were put in fresh mortar, then removed after a specific time of exposure, and finally tested in tension to quantify the strength decay. Two different mortars were employed:

- a hydraulic lime-based mortar: a structural grout of natural hydraulic lime for impregnation of fabrics for structural reinforcement provided by the company Innovation s.r.l. with the name of *FIDCALX NHL5*®;
- a cement-based mortar: a mortar based on hydraulic cement binders for bonding and embedment of fabrics (strengthening composite systems) provided by the company Innovation s.r.l. with the name of *Kimisteel LM Sta-0217*.

The selected mortars are commonly adopted as matrix in fabric reinforced composite systems. For both the environments 3 series of flax threads were studied differing in the duration of exposure, notably of 7, 28 and 56 days. Such method allows to correlate the degradation phenomena with time of exposure. As defined in the technical sheet provided by the manufacturer, the hydraulic lime- and the cement-based mortars respectively presented a compressive strength of 15 MPa and 45 MPa. The ageing protocol comprises the follow environment conditions:

- *Flax-T H-lime (7-28-56 days)*: the threads were embedded in hydraulic lime mortar for a period of 7, 28 and 56 days;
- *Flax-T Cement (7-28-56 days)*: the threads were embedded in cement mortar for a period of 7, 28 and 56 days.

The ageing was carried out at laboratory temperature, 20 °C, and relative humidity, 50%.

3.3.1.3 Ageing Protocol 3

The ageing protocol 3 comprises water solutions defined according to the literature [8] to reproduce hydraulic lime- and cement-based mortar environments. Two different water solutions, differing in the chemical components adopted, were considered. The tests were conducted at controlled temperature to reproduce an accelerated ageing process. The ageing protocol comprises the environment conditions described as follow:

- *Flax-T Environment A (7-28-56 days)*: the threads were immersed for periods of 7, 28 and 56 days in solution meant to reproduce hydraulic lime conditions (16 wt% of $Ca(OH)_2$ in distilled water, pH = 12.37). A closed container with fibres immersed in the solution was kept at 55 °C;
- *Flax-T Environment B (7-28-56 days)*: the threads were immersed for periods of 7, 28 and 56 days in a solution meant to reproduce cement conditions (16 wt% of $Ca(OH)_2$, 1% of Na(OH) and 1.4% of K(OH) in distilled water, pH = 13.1). A closed container with fibres immersed in the solution was kept at 55 °C.

This ageing protocol was conceived to reproduce the real decay of mechanical properties of the fibres in the long term once embedded in inorganic matrices.

3.3.2 Results and Discussion

Figure 3.11 shows the stress-strain response of the series of specimens subjected to the ageing protocol 1. Specifically, the three graphs concern the threads immersed for 1000 h respectively in water, saltwater and alkaline solution. The threads were extracted from the fabric in the warp direction, therefore, the series Flax-T-warp was considered as a control series representing the response before the ageing process. The main values of the mechanical properties, with the respective coefficient of variation, are listed in Table 3.3. According to the ageing protocol 2 samples embedded either in a cement- or in a hydraulic lime-based mortar were tested after a conditioning period of 7, 28, and 56 days. The tensile response, in terms of stress-strain,

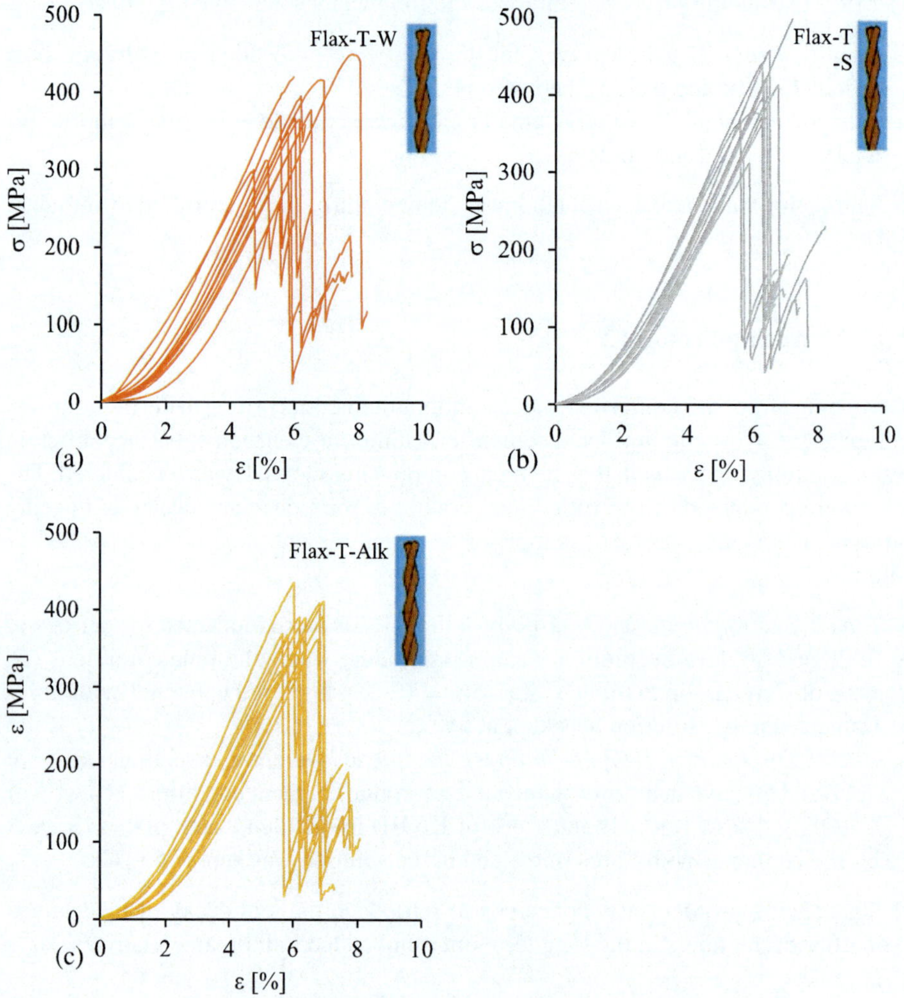

Fig. 3.11 Tensile stress-strain diagrams of flax threads subjected to ageing protocol 1

Table 3.3 Mechanical properties of flax samples subjected to ageing protocol 1

Ageing environment		Curing period	Numb. of spec.	P_u		f_t		ε_u	
				Mean (N)	Co.V. (%)	Mean (MPa)	Co.V. (%)	Mean (%)	Co.V. (%)
Flax-T	Control	–	12	92	10	361	10	5.20	11
Flax-T	Water	1000 h	12	96	11	377	11	6.08	13
Flax-T	Saltwater	1000 h	13	102	13	401	13	6.29	7
Flax-T	Alkaline water	1000 h	13	96	9	376	9	6.34	7

for each conditioning environment and for each conditioning period, is shown in Figs. 3.12 and 3.13. The main values of the mechanical properties, with the respective coefficient of variation, are listed in Table 3.4.

The results of stress strain curves concerning the specimens subjected to the ageing protocol 3 are shown in Figs. 3.14 and 3.15. Tensile tests were respectively performed after 7, 28 and 56 days. Due to the high degradation of the specimens subjected to the Environment A for a conditioning period of 56 days, no tensile response was recorded concerning this series. The main values of the mechanical properties, with the respective coefficient of variation, concerning the samples aged in Environment A and B are listed in Table 3.5.

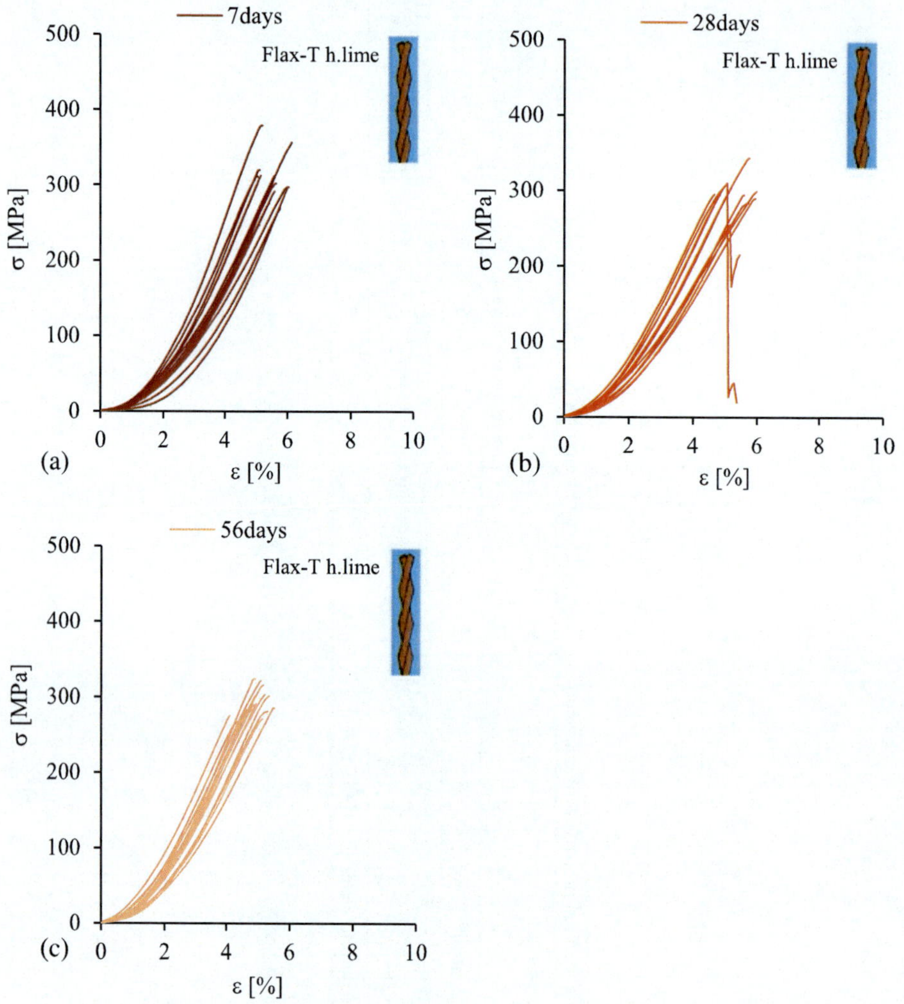

Fig. 3.12 Stress-strain curves of threads in hydraulic lime mortar (ageing protocol 2)

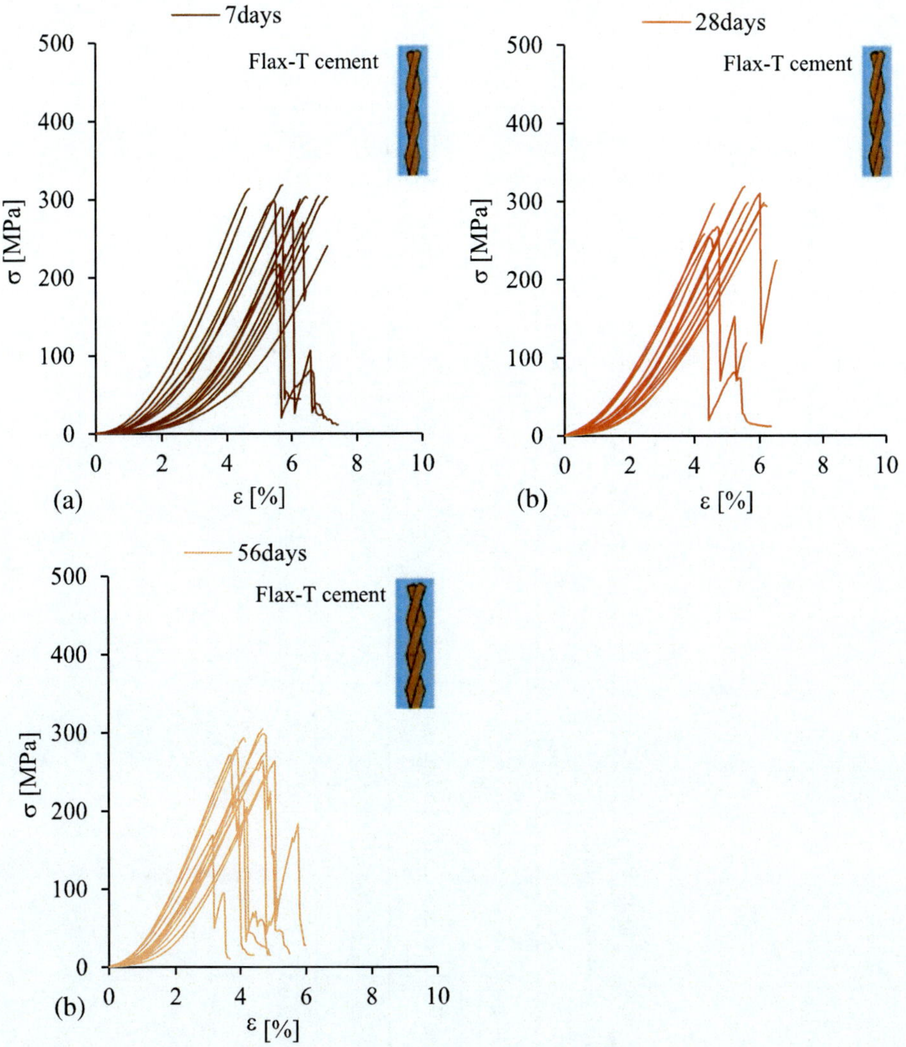

Fig. 3.13 Stress-strain curves of threads in cementitious mortar (ageing protocol 2)

Some of the series are characterised by less than 15 samples. Few specimens were discarded due to a not correct execution of the test. Tensile strength, f_t, was calculated based on the nominal cross-section of Flax-T specimens in the unexposed configuration. The ageing protocol 1 is a preliminary analysis in view of the application of the fabric in inorganic matrix-based composites defined according to acceptance criteria (ACI-434, 2018) (Fig. 3.16). Figure 3.17 shows no significant reduction of the tensile strength of the exposed specimens with respect to the unexposed reference ones. This result is in line with a similar study from the literature in which the slight reduction of strength was attributed to statistic causes [13].

Table 3.4 Mechanical properties of flax samples subjected to ageing protocol 2

Ageing environment		Curing period	Numb. of spec.	P_u		f_t		ε_u	
				Mean (N)	Co.V. (%)	Mean (MPa)	Co.V. (%)	Mean (%)	Co.V. (%)
Flax-T	Control	–	12	92	10	361	10	5.20	11
Flax-T	Hydraulic lime mortar	7 days	15	78	10	306	10	3.45	7
Flax-T		28 days	12	75	8	295	8	3.68	9
Flax-T		56 days	15	75	6	293	10	3.44	9
Flax-T	Cementitious mortar	7 days	15	72	10	284	10	3.34	9
Flax-T		28 days	14	71	10	280	10	3.45	12
Flax-T		56 days	15	67	13	262	13	3.04	11

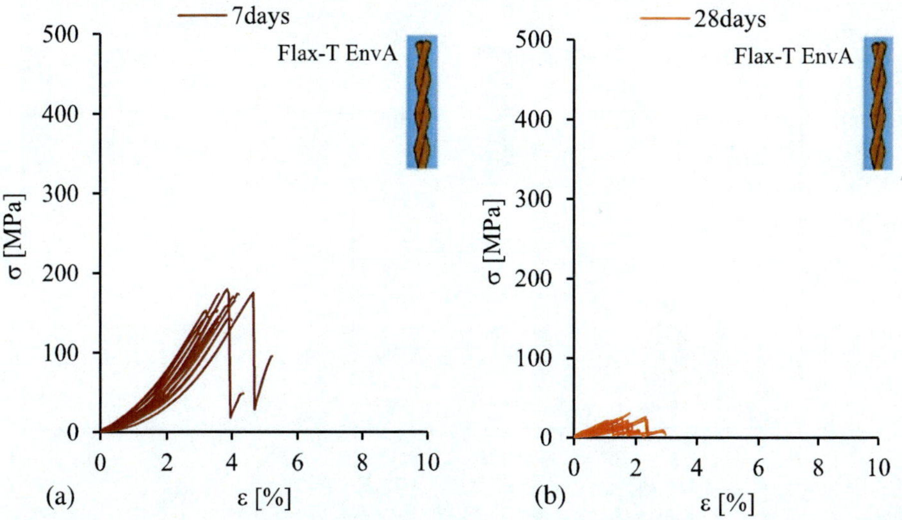

Fig. 3.14 Stress-strain diagram of flax threads in Environment A (ageing protocol 3)

An appreciable increase of deformability was indeed observed with larger main values of the ultimate strain, ε_u. This experimental evidence may be attributed to some material degradation (weight loss) leading to the loosening of the twisted yarn, and, hence, to larger strain values.

The experimental evidence highlights the ineffectiveness of the adopted ageing conditions once applied on the fabric, not enough aggressive to trigger any ageing phenomena. A more aggressive ageing protocol, perhaps by considering longer time exposures and/or higher temperatures, may enhance the efficiency of the durability test making possible to analyse actual degradation phenomena.

Figure 3.18 shows the mean tensile strength concerning the series subjected to the ageing protocol 2. A moderate decrease of strength was shown in both hydraulic lime and cement matrices conditioning environments. In both the cases the decreasing deterioration over time seems to mainly occur in the first days of exposure, together with the mortar hydration process, and then to be less significant in the next period until 56 days. This strength reduction is equal to about the 16% for the hydraulic lime-based environment, and of about the 22% in the case of the cement-based conditions. Besides the short exposure time, it is confirmed the higher aggressivity of cement mortars with respect to the lime-based ones, in line with the literature [6] (Fig. 3.19).

It is worth to highlight that such slight decay of strength may also be partially attributed to a superficial damage of the bundle as a consequence of an abrasive effect when fibres were extracted from the hardened mortar prior the testing. Further investigation, complying with long term exposures conditions, are needed to deeply define the ageing of flax fibres due to the aggressiveness of the mortar hydration products. A more detailed investigation of the degradation phenomena on the external surfaces of the flax filaments was carried out via a Scanning Electron Microscope

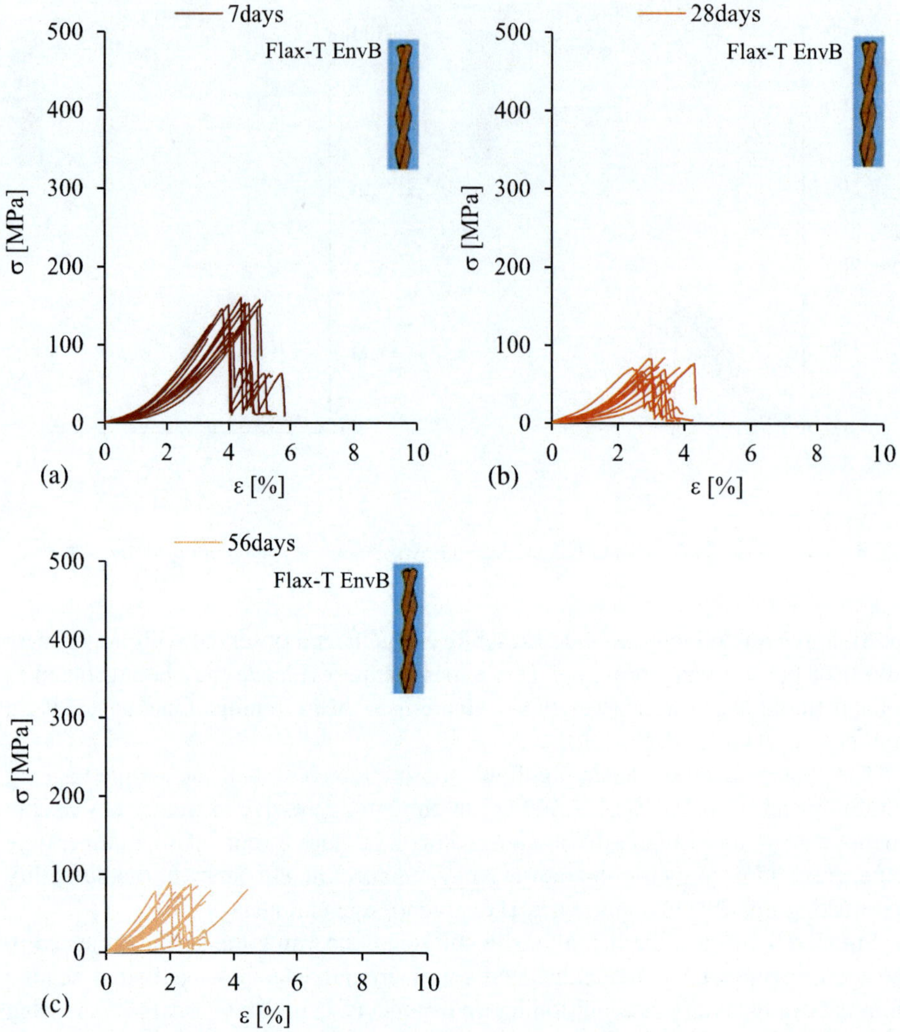

Fig. 3.15 Stress-strain diagram of flax threads in Environment B (ageing protocol 3)

analysis. Filaments comprising flax threads exposed for 28 days in hydraulic lime and cement mortars are shown in Figs. 3.20 and 3.21. A reference unexposed sample is shown in Fig. 3.22 for comparison.

The widespread presence of hydration products on the surface of the flax filaments emphasises a good bond interaction between fibres and matrix. The filament's external wall results clearly damaged with respect to the reference one, due to the alkaline hydrolysis process. This phenomenon causes, on a global scale, the reduction of the aged thread tensile strength.

Table 3.5 Mechanical properties of flax samples subjected to ageing protocol 3

Ageing environment		Curing period	Numb. of spec.	P_u		f_t		ε_u	
				Mean (N)	Co.V. (%)	Mean (MPa)	Co.V. (%)	Mean (%)	Co.V. (%)
Flax-T	Control	–	12	92	10	361	10	5.20	11
Flax-T	Environment A	7 days	15	39	12	151	12	2.48	8
Flax-T		28 days	12	5	10	21	10	1.18	9
Flax-T		56 days	–	–	–	–	–	–	–
Flax-T	Environment B	7 days	15	36	11	141	11	2.69	11
Flax-T		28 days	14	18	12	70	12	1.89	8
Flax-T		56 days	15	19	12	76	10	1.66	9

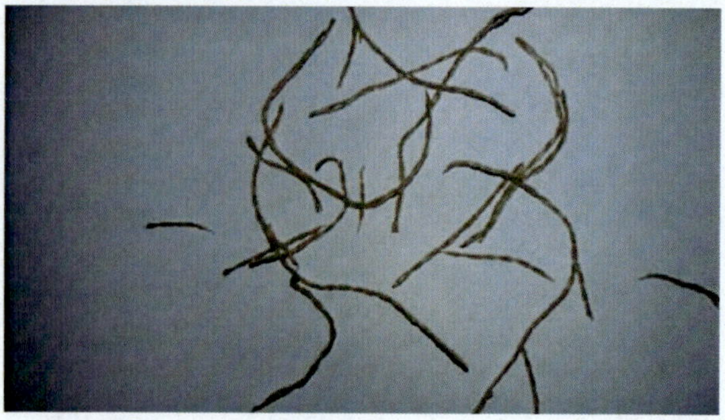

Fig. 3.16 Degraded Flax-T Env A threads after 56 days of exposure

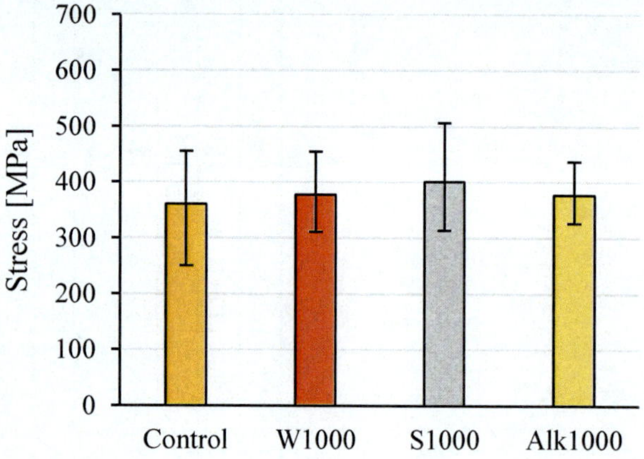

Fig. 3.17 Tensile strength of series belonging to the ageing protocol 1

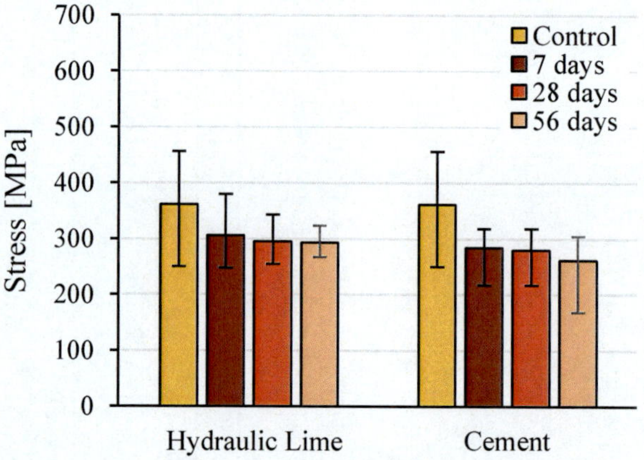

Fig. 3.18 Tensile strength of series belonging to the ageing protocol 2

However, the bond between fibre and matrix in composite systems is a crucial aspect both in case of synthetic and plant fibres [2, 4]. Thus, even if the tensile strength decreases, an increase of the fibre surface roughness may improve the adhesion with the matrix, with consequent beneficial effects in the entire composite mechanical response.

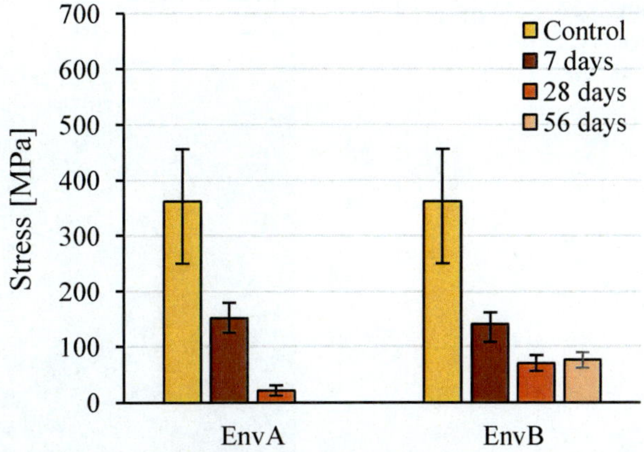

Fig. 3.19 Tensile strength of series belonging to the ageing protocol 3

Fig. 3.20 SEM image of Flax-T h.lime 28 days sample (5000X)

Fig. 3.21 SEM image of Flax-T cement 28 days sample (5000X)

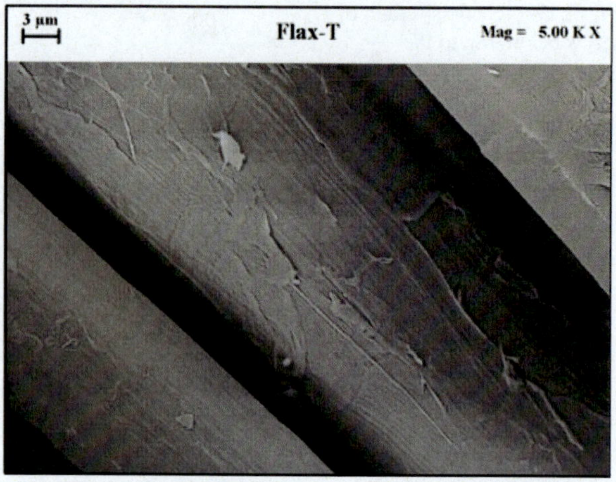

Fig. 3.22 SEM image of a reference Flax-T sample (5000X)

The ageing protocol 3 comprises two water solutions, specifically conceived to reproduce un accelerated degradation of the fibre in hydraulic lime and cement mortar environments. High temperatures accelerate the ageing process allowing to simulate the degradation of the fibres in the long-term. However, it rises uncertainties related to the corresponding time in natural temperature conditions.

The time of accelerated exposures of 7, 28, and 56 days at a temperature of 55 °C, should correspond respectively to 3, 11 and 20 years in natural conditions according to Micelli and Aiello (2019). Figure 3.19 show the main value of the tensile strength of the specimens subjected to the ageing protocol 3. Both the ageing Environments A and B led to a decay of the strength of about the 60% after 7 days of conditioning. Flax threads exposed to the Environment A showed a loss of strength of the 95% after 28 days, and completely disintegrated after 56 days (Fig. 3.16). Flax threads aged in the Environment B exhibited a decay of strength of about the 80% after 28 days, and a quite constant value of the strength in the next period until 56 days. The alkaline hydrolysis of lignin and non-cellulosic constituents of flax filaments, accelerated by the high temperature of 55 °C, may be responsible of the degradation of the fibres.

SEM images reporting flax samples aged for 28 days in Environment A and B are respectively shown in Figs. 3.23 and 3.24. Figure 3.25 shows a SEM image concerning a non-exposed a reference sample. The specimen aged in the Environment A shows a higher number of impurities, waxes and non-cellulosic compound removal on the filament external surface, and a larger amount of degradation products mixed to solution residues (Fig. 3.23) with respect to the specimen aged in Environment B (Fig. 3.24). As matter of fact, Environment A resulted more aggressive than Environment B.

The higher strength of specimens exposed to the Environment B may be due to an increase of cellulose crystallinity: according to literature [2] the removal of non-cellulosic materials due to the high alkalinity, may cause a better packing of cellulose chains with a consequent increase of the crystallinity index.

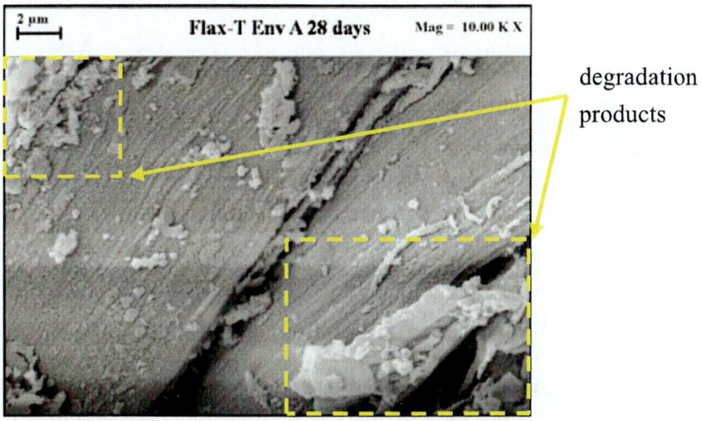

Fig. 3.23 SEM image of Flax-T Env-A 28 days sample (10000X)

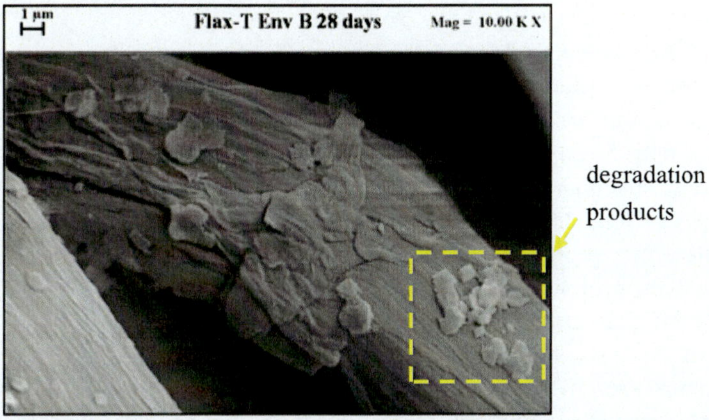

Fig. 3.24 SEM image of Flax-T Env-B 28 days sample (10000X)

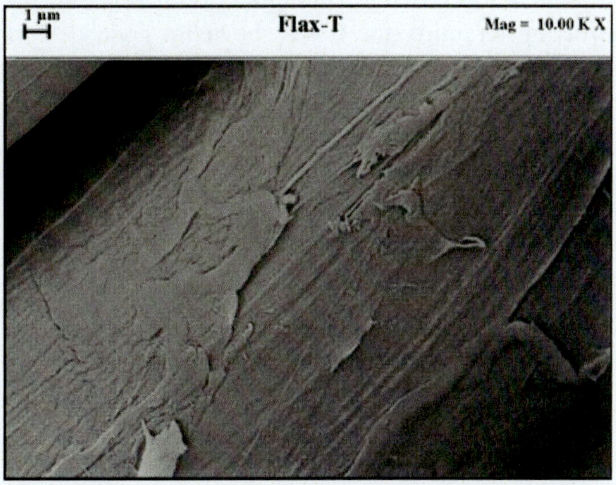

Fig. 3.25 SEM image of a reference Flax-T sample (10000X)

3.4 Conclusions

The present chapter reports the mechanical properties in the short and long terms of a Flax textile to be used as reinforcement in Textile Reinforced Mortar (TRM) composite systems. The main findings of the experimental analysis are summarised as follows:

- the mean value of the tensile strength of the flax mean thread was 354 MPa, the Young Modulus was 9.3 GPa and the strain at failure was 5.6%. This mechanical behaviour is consistent with results reported in literature concerning similar plant fibres [9];
- the decay of the mechanical behaviour in the long terms, induced by the adopted ageing protocols, was lower than expected;
- notably, specimens exposed to the ageing protocol 1, including immersion in water, saltwater and alkali solution for 1000 h, did not present significant loss of strength;
- similarly, flax threads subjected to protocol 2, immersed in hydraulic lime and cement mortar, suffered a decrease of strength of about 16 and 22% after 7 days. However, no further decrease of strength was exhibited for longer immersion durations until 56 days;
- conversely, specimens exposed to the ageing protocol 3, immersed in alkaline solutions at a temperature of 55 °C, showed a significant loss of strength with a complete degradation of the flax threads in specimens immersed for 56 days. This was due to the prolonged exposure at harsh ageing conditions, in terms of pH and temperature, that promoted flax fibre degradation;
- finally, the analysis of SEM images proved a good adhesion between fibres and the adopted matrices (hydraulic lime- and cement-based mortars) due to interlocking positions created onto fibres surface as result of alkaline hydrolysis.

The results show that the flax textile exhibits valuable mechanical properties and represents a promising solution as reinforcement in TRM composite systems. In terms of durability the fibres showed a good performance in the short term, and significant decreases of strength in accelerated ageing exposures. This behaviour can be extended to that of the fabric adopted as reinforcement in TRMs.

Further investigations are needed, perhaps by focusing on fabric treatments and/or coating, with the aim of improving both the mechanical behaviour and the long-terms performance. Moreover, the mechanical and durability investigation needs to be extended to the Flax TRM systems as whole to confirm the potential of this kind of sustainable composite system.

References

1. AC 434-1011-R1 (2018) Acceptance Criteria for Masonry and Concrete Strengthening using Fiber-Reinforced Cementitious Matrix (FRCM) Composite Systems, ICC Evaluation Service, LLC: Birmingham, AL, USA
2. Aly M, Hashmi MSJ, Olabi AG, Benyounis KY, Messeiry M, Hussain AI, Abadir EF (2012) Optimization of alkaline treatment conditions of flax fiber using box-Behnken method. J Nat Fibers 9(4):256–276
3. Ardanuy M, Claramunt J, Toledo Filho RD (2015) Cellulosic fiber reinforced cement-based composites: a review of recent research. Constr Build Mater 79:115–128
4. Cevallos OA, Olivito RS (2015) Effects of fabric parameters on the tensile behaviour of sustainable cementitious composites. Composites Part B Engineering, 69:256–266
5. Coppola B, Di Maio L, Scarfato P, Incarnato L (2015) Use of polypropylene fibers coated with nano-silica particles into a cementitious mortar. AIP Conf Proc 1695(1):020056
4. Ferrara G, Coppola B, Di Maio L, Incarnato L, Martinelli E (2019) Tensile strength of flax fabrics to be used as reinforcement in cement-based composites: experimental tests under different environmental exposures. Compos Part B Eng 168:511–523
5. Ferrara G, Pepe M, Martinelli E, Toledo Filho RD (2019) Influence of an impregnation treatment on the morphology and mechanical behaviour of flax yarns embedded in hydraulic lime mortar. Fibers 7(4):30
6. Fidelis MEA, Toledo Filho RD, de Silva AF, Mechtcherine V, Butler M, Hempel S The effect of accelerated aging on the interface of jute textile reinforced concrete. Cement Concrete Compos 74:7–15
7. ISO 13934-1 (2013) Textiles-tensile properties of fabrics- Determination of maximum force and elongation at maximum force using the strip method
8. Micelli F, Aiello MA (2019) Residual tensile strength of dry and impregnated reinforcement fibres after exposure to alkaline environments. Composites Part B Engineering, 159:490–501
9. Mustata A, Mustata FSC (2013) Moisture absorption and desorption in flax and hemo fibres and yarns. Fibres Text Eastern Europe 31, 3(99):26–30
10. Nobili A (2016) Durability assessment of impregnated Glass Fabric Reinforced Cementitious Matrix (GFRCM) composites in the alkaline and saline environments. Construction and Building Materials, 105:465–471
11. Olivito RS, Cevallos OA, Carrozzini A (2014) Development of durable cementitious composites using sisal and flax fabrics for reinforcement of masonry structures. Materials and Design, 57:258–268
12. Ramamurthi DS, Sophia M (2016) A review on modified lime-based mortars – an alternative to cement mortar. International Journal of Innovative Science, Engineering and Technology 2(12):350–356
13. Sathishkumar TP, Navaneethakrishnan P, Shankar S, Rajasekar R, Rajini N (2013) Characterization of natural fiber and composites—a review. J Reinforced Plastics Compos
14. Technical Sheet (2019) FIDFLAX GRID 300 HS20®, Tessuto a Rete Bilanciato in Fibra di Lino per il Rinforzo Strutturale. Available online: http://www.fidiaglobalservice.com. Accessed 29 Dec 2019
15. Wei J, Ma S, Thomas DG (2016) Correlation between hydration of cement and durability of natural fiber-reinforced cement composites. Corros Sci 106:1–15
16. Yan L, Kasal B, Huang L (2016) A review of recent research on the use of cellulosic fibres, their fibre fabric reinforced cementitious, geo-polymer and polymer composites in civil engineering. Composites Part B Engineering 92:94–132

Chapter 4
Mechanical Behaviour and Durability of Flax TRMs: Tensile and Shear Bond Tests

Once defined the properties of the TRM constituent materials, i.e. mortar and textile, to get an exhaustive analysis it is necessary to move from a "*material*" scale of analysis to the investigation of the composite itself. As widely discussed in Chap. 2, in order to qualify and characterise a TRM composite, it is necessary to identify its mechanical behaviour in tension and its interaction with a masonry substrate. With the aim of providing the mechanical response of the studied Flax TRM system, an experimental study is proposed at the "*composite*" scale of analysis. Specifically, the results of tensile, carried out on Flax TRM coupons, and of shear bond tests on Flax TRM-to-masonry substrate elements, are presented in this section. In addition, in view of a more comprehensive characterisation, and in line with what is proposed by standards and guidelines, the results of an experimental campaign aimed at investigating the mechanical performance of Flax TRMs exposed to several ageing environments, is presented as well.

4.1 Tensile Test on Flax TRM Systems

4.1.1 Materials

The mortar adopted for the implementation of the Flax TRM system was hydraulic lime based. Different studies showed that the use of mortars comprising lime binders is an efficient solution to implement composite systems for the strengthening of historical masonry structures [18]. The pre-mixed mortar adopted in the study is provided by the Italian company "Innovations s.r.l." [7] with the name of [23]. The mix design comprises pure natural hydraulic lime binder and selected aggregates with maximum granulometry of 2.5 mm. The inert materials size distribution is shown in Fig. 4.1. The part of the curve concerning the grains bigger than 0.074 mm was obtained by means of a vibrating test sieve machine [3], the rest of the curve concerning the finest particles was obtained by means of a laser diffraction particle

G. Ferrara, *Flax-TRM Composite Systems for Strengthening of Masonry*, Springer Theses,
https://doi.org/10.1007/978-3-030-70273-1_4

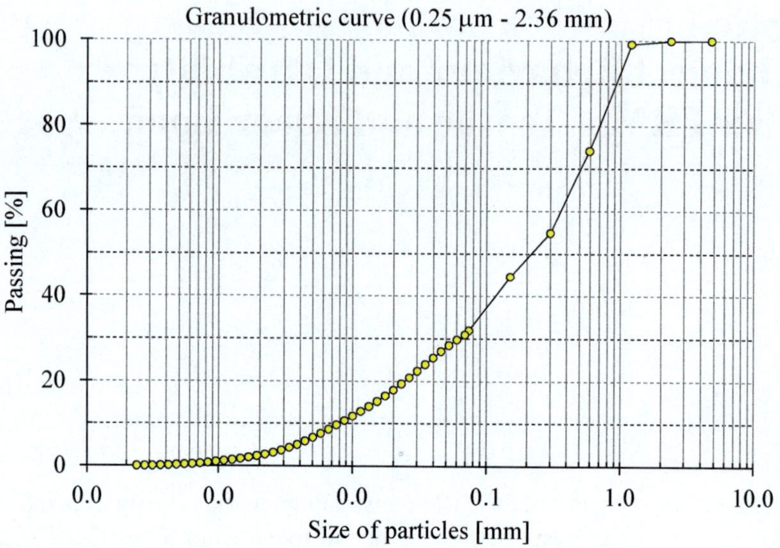

Fig. 4.1 Grain size distribution curve for lime-based premixed mortar powder

sizing machine. As known, in TRM implementation, the mortar penetrability within the textile is hindered if the maximum aggregate diameter is about 2 ÷ 2.5 times larger than the fabric clear opening. Being the 99% of the aggregates smaller than 1.19 mm and being the flax textile clear opening of about 3 mm, it can be considered a compatibility between the two composite components.

According to manufacturer instructions, the mix-design included an amount of water of 19% (in weight) of the pre-mixed formula (binder + aggregates). Consequently, the mixture had a water-to-binder ratio of 0.60. The consistency of the mortar at the fresh state, assessed according to EN 1015-3 [10], was 230 mm. Flexural and compressive strength of the hardened mortar (at 28 days) were assessed for each mortar batch employed in TRMs casting, on prismatic samples performed according to EN 196-1 [11]. The mortar exhibited mean flexural and compressive strength respectively equal to 3.58 MPa and 8.02 MPa.

The textile adopted consists in the flax fabric described in Chap. 3, whose physical properties are listed in Table 4.1 and whose mean mechanical properties, assessed

Table 4.1 Main properties of the tensile test series of specimens

Series	Number of specimens	Section (mm x mm)	A_{mortar} (mm^2)	$A_{textile}$ (mm^2)	vf (%)	$f_{textile}$ (MPa)
Flax TRM-1L-T	4	5 × 60	300	6.1	2.04	286
Flax TRM-2L-T	5	8 × 60	480	12.2	2.55	

on a single thread, are a tensile strength of 354 MPa, a Young's modulus of 9.3 GPa and a strain at failure of 5.6% [13].

4.1.2 Methods

Flax TRM coupons were performed, by means of a wet lay-up method, according to the guidelines reported in the ACI 549.4R-13 [2]. The specimens were cast by alternating mortar layers and flax textile plies. The base of the mould was filled with the first layer of mortar for a depth of 2.5 mm, then the flax strip was placed on it by ensuring it to be clamped along two edges of the specimens, so that it was kept as stretched as possible (Fig. 4.2a). A roller was adopted to press the textile for guaranteeing a full impregnation within the mortar (Fig. 4.2c). The next layer of mortar, having the same thickness of the first one, was placed on it. The procedure was iterated for multi-plies' configurations in which more than one textile layer were placed within the system, always making sure a layer of mortar to be placed between two fibre strips to completely fill the grid voids. The application of the outer layer of mortar was performed by thoroughly smoothing it (Fig. 4.2b).

Two reinforcement configurations were taken into account in the present study: the first one, labelled as Flax-TRM-1L, consisted of a TRM characterised by a thickness of 5 mm and by one layer of reinforcing flax textile, with a fibre volumetric ratio equal to 2.04%; the second one, labelled as Flax-TRM-2L, was characterised by a thickness of 8 mm and by two plies of flax fabric overlapping each other by means of a thin layer of mortar, with a resulting fibre volumetric ratio equal to 2.55%.

Two series of specimens, performed by means of the two above mentioned reinforcement configurations, were considered:

- Flax-TRM-1L-T: specimens characterised by the reinforcement configuration Flax-TRM-1L, and tested in tension (Fig. 4.3a);

Fig. 4.2 Flax TRM specimen implementation: **a** Application of the textile; **b** Application of the outer layer of mortar; **c** Pressure application on the textile

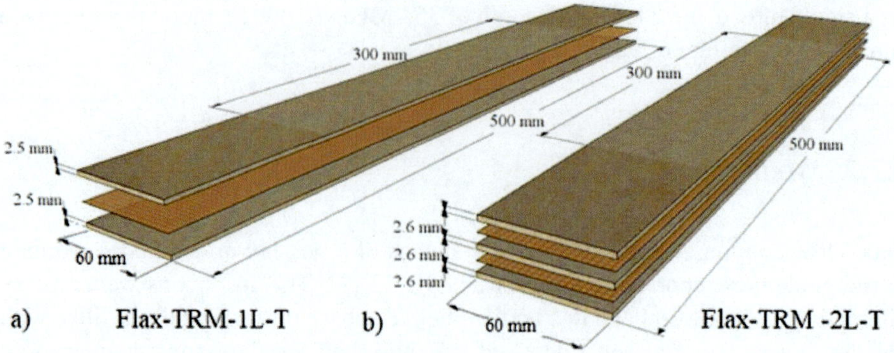

Fig. 4.3 Flax-TRM coupons tested in tension: **a** Flax-TRM-1L-T, comprising one layer of flax textile; Flax-TRM-2L-T, comprising two layers of flax textile

- Flax-TRM-2L-T: specimens characterised by the reinforcement configuration Flax-TRM-2L, and tested in tension (Fig. 4.3b).

The specimens were characterised by a length of 500 mm, of which 300 mm in the middle represented the free length between the steel tabs glued to the specimen edges to transfer the load from the testing machine to the TRM. Further details concerning the two series of specimens are shown in Table 4.1, in which f_{textile} represents the main tensile strength exhibited by flax strips 60 mm large (series labelled as Flax fabric-6 cm in Chap. 3), and vf the fibre volumetric reinforcement ratio. The coupons, characterised by a width of 60 mm (Fig. 4.4a), contained, with respect respectively to the Flax-TRM-1L-T and Flax-TRM-2L-T series, 24 and 48 flax threads (the thread to which reference is made is the one defined as Flax-T in Chap. 3). For each series of specimens three mortar prisms were cast to mechanically characterise the matrix (Fig. 4.4b).

Tensile tests were performed at the STRENGTH laboratory (University of Salerno) via a Zwick Roell Schenck Hydropuls S56 machine, with 630 kN maximum capacity (Fig. 4.5b). During the test, carried out in displacement control with a rate of

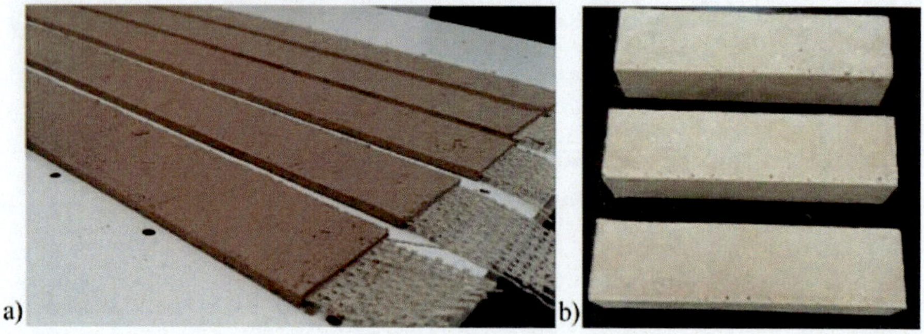

Fig. 4.4 **a** Flax-TRM-1L-T specimens; **b** Mortar prisms

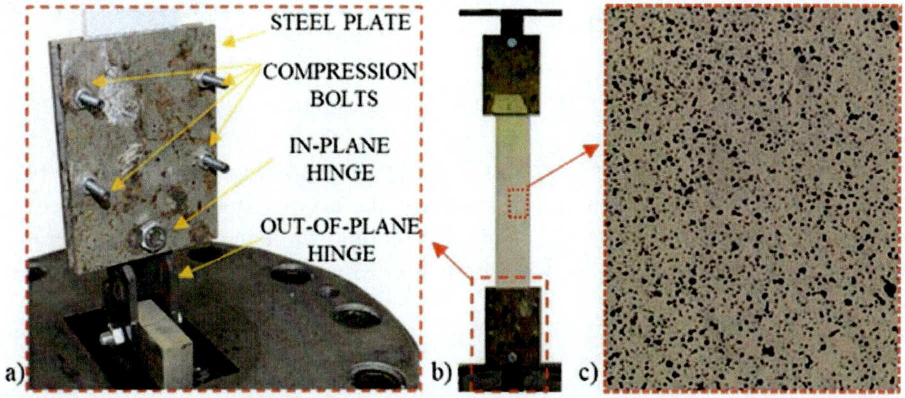

Fig. 4.5 **a** TRM tensile test clamping system; **b** TRM tensile test set-up; **c** surface treatment for digital image correlation photos

0.25 mm/min, the applied load and the displacement between the clamping wedges were recorded.

The specimens were gripped at the two edges for a length of 100 mm, by means of a steel system (Fig. 4.5a). Tensile load was transferred from the stiff steel plates, glued with an epoxy resin, to the specimen edges by shear stresses over a bond length of 100 mm (clevis system). If needed, compression can be applied on the specimen's edges, by tightening four bolts placed at the corners of the plates (clamping system). A rotational release in the specimen plane was created by means of a bolt placed in the middle lower side of the plate (in-plane hinge). This bolt connected the gripping system to a metallic device that guaranteed the out of plane rotations (out-of-plane hinge). The entire system was properly conceived to avoid bending and twisting moments actions on the specimen. A fully clamped configuration can be created with the locking of the hinges by tightening the bolts. In the present study a clevis configuration was reproduced, and so no compression was applied at the specimens' edges and both rotations were permitted.

In order to carry out Digital Image Correlation (DIC) analysis, the outer surface of the specimen was properly treated along the gauge length. The treatment consisted in creating a white background on which black dots were stochastically sprinkled by means of a spray can (Fig. 4.5c). DIC analysis was proved as an efficient technique to obtain the displacement field of TRMs tested in tension [4, 22]. The equipment for the DIC analysis comprised a camera placed at a focal distance of 500 mm. Pictures were acquired with a rate of 3 ph/min, with a resolution of 3454 $\times$ 5184 pixels (0.06 mm/pix).

4.1.3 Results and Discussion

The tensile response of Flax TRMs in terms of load-displacement is shown in Figs. 4.6 and 4.7.

The tensile response in terms of stress-strain is shown in Figs. 4.8 and 4.9. When the specimen is uncracked, it can be assumed the normal stress to be uniformly distributed through the entire composite cross section, while in proximity of a crack the entire applied load can be assumed to be borne by the dry textile. Therefore, the

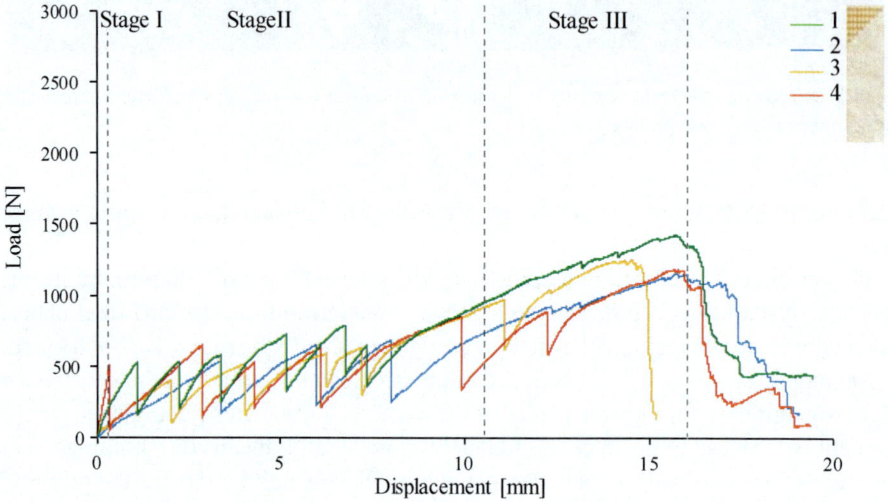

Fig. 4.6 Load-displacement curves concerning Flax TRM-1L-T specimens

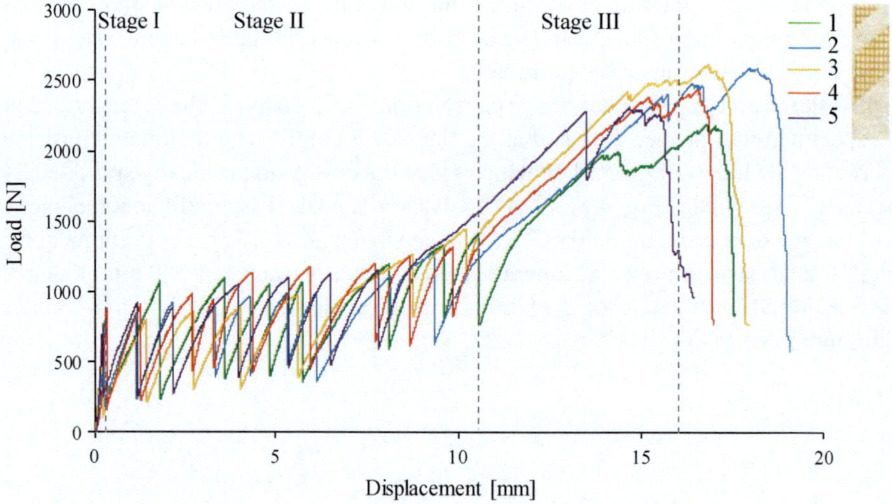

Fig. 4.7 Load-displacement curves concerning Flax TRM-2L-T specimens

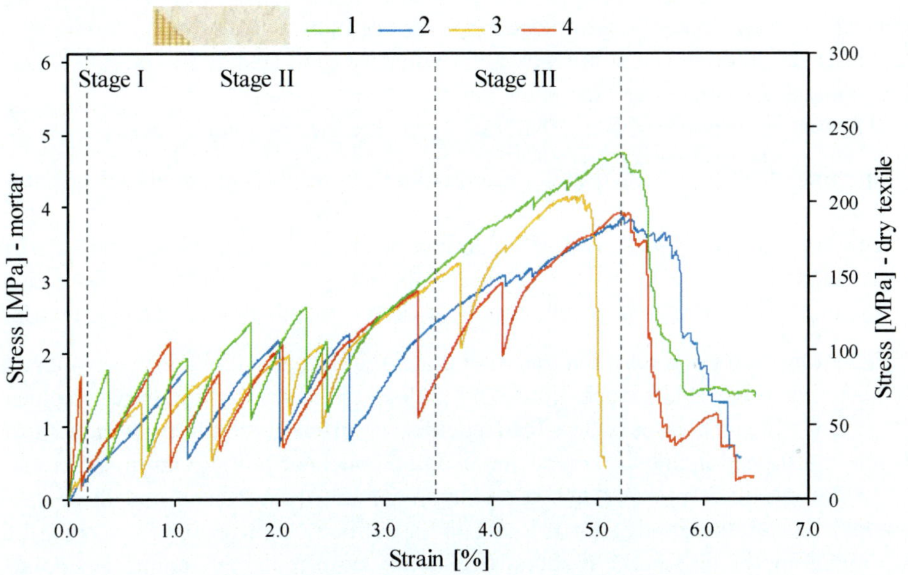

Fig. 4.8 Stress-strain curves concerning Flax TRM-1L-T specimens

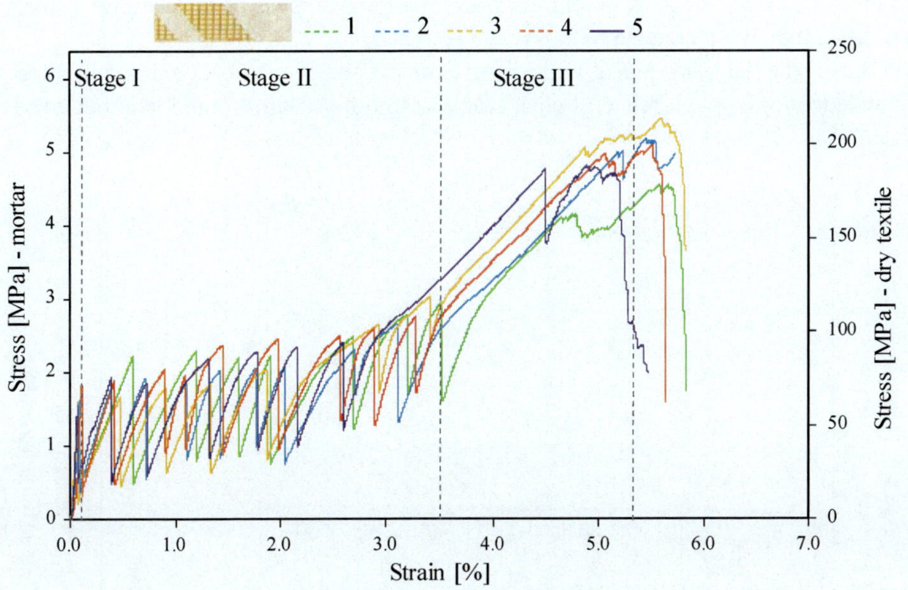

Fig. 4.9 Stress-strain curves concerning Flax TRM-2L-T specimens

normal stress was assessed with respect to both the mortar cross section and the dry textile cross section. The strain was approximately evaluated as the ratio between the global displacement and the gauge length.

The tensile response of Flax TRMs was characterised by three stages:

- the first stage represented the elastic response up to the occurrence of the first crack;
- the second one represented a transition behaviour in which several cracks developed;
- the third one in which the mechanical behaviour is mainly governed by the textile.

The transition point between the second and the third stage (σ_2, ε_2) corresponds to the point, after the last crack, in which the slope of the stress-strain curve changes (Fig. 4.10). In some cases, where the last crack occurred in proximity of the failure mode, this transition point is considered in correspondence of the penultimate crack.

In the elastic response stage, the coupon remained uncracked and the entire composite bore the applied load. Being the specimen characterised by a constant section along the length, the first crack randomly occurred at any point, depending on geometric imperfections, or concentrations of stress in proximity of the loaded edges. Except for the specimen Flax TRM-1L-T-4, the other specimens belonging to the series Flax TRM-1L-T did not present an elastic phase because, due to their thinness, they were cracked before being tested.

During the transition phase, large displacement values were observed and, in correspondence to the occurrence of each crack, a significant drop in the load occurred.

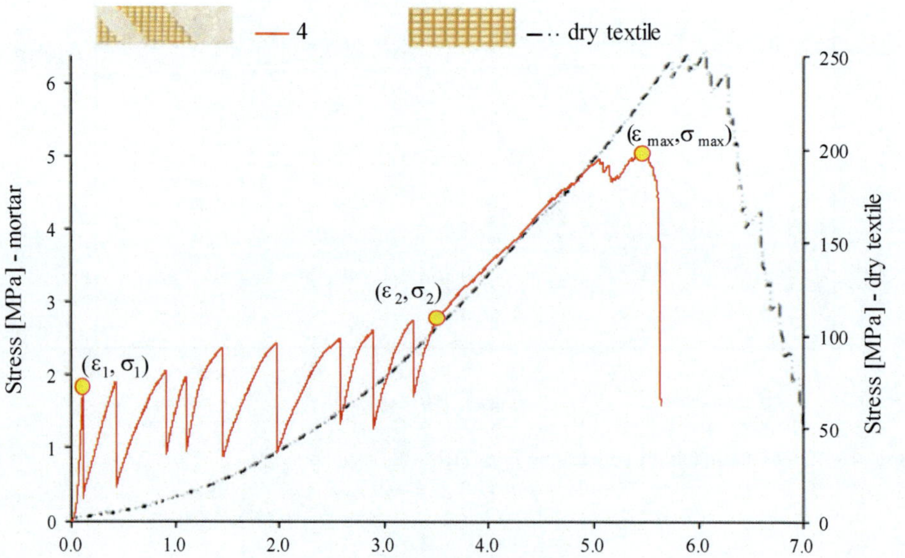

Fig. 4.10 Comparison between stress-strain response in tension of a representative TRM specimen (series Flax TRM-2L-T), and a representative dry textile (series Flax fabric-6 cm)

Such mechanical response, in line with similar research studies in which plant fibres were employed [5, 12, 21], resulted not consistent with the typical behaviour exhibited by TRM composite with high strength synthetic textiles. Moreover, such behaviour may be also due to a not adequate degree of mechanical interlock between matrix and reinforcement. This issue may be solved by correctly defining the matrix mix design (aggregates granulometry and consistency).

Such behaviour may be attributed to two aspects. On one hand, the (out of plane) deformability of the textile, due to the absence of any impregnation treatment, unavoidably resulted in the textile not to be uniformly stretched during the specimens' implementation, even though great attention was devoted to this aspect during the manufacturing process. The latter is a common aspect also in the conventional TRM coupons when dry textile is adopted. On the other hand, as widely discussed in Chap. 3, the mechanical response of plant fibre textile loaded in tension is typically characterised by a lower stiff initial response, due to specific macro- and micro-structure of the threads. As a result, in the elastic behaviour, at low strain levels, the textile was only partially activated, while the load is almost entirely borne by the mortar. Indeed, in terms of stress, the σ_1 is approximately equal to 50% of the mortar flexural strength (being the latter un overestimation of the uniaxial direct tensile strength of the mortar). Consequently, at the occurrence of the first crack, leading to partial loss of the contribution of the matrix, the lower contribution of the textile caused significant drops of the load. To qualitatively deeper understand such behaviour, in Fig. 4.10, the stress-strain response of a representative specimen (Flax TRM-2L-T4) was compared with the response of a representative flax strip tested in tension (series labelled Flax fabric-6 cm in Chap. 3). Such comparison, although being approximated due to the different test conditions of the two specimens compared (the flax strips were characterised by a free length of 15 cm and were clamped; the textile within the TRM, in proximity of a crack, was under a significant concentration of stress over a short length), emphasises the discrepancy in terms of stiffness between the TRM and the dry textile at a low strain level. The comparison also highlights a similarity between the TRM stiffness in the third stage, and the dry textile stiffness in the final linear branch, confirming that, once crack saturation has been achieved, the response is mainly governed by the dry textile.

The main mechanical parameters deriving from tensile tests on the two series of specimens, together with the mean value and the coefficient of variation for each series, are listed in Table 4.2. Specifically, P_{max} represents the maximum load, $disp_{max}$ represents the displacement corresponding to the maximum load, σ_{max} represents the maximum tensile stress assessed with respect to the entire TRM cross section, ε_{max} represents the axial strain corresponding to the maximum stress, $\sigma_{textile}$ represents the maximum tensile stress assessed with respect to the flax textile cross section, $\sigma_{max}/f_{textile}$ represents the exploitation ratio evaluated as the ratio between the maximum textile stress achieved during the test and the textile strength. The failure mode is reported as well and named as A, when the fibres failed in a section in proximity to the end-fixtures, with B in the case of a random section through the length. As expected, the results from the two series show that by doubling the amount of textile, the maximum capacity in terms of load nearly doubled. Such experimental evidence

Table 4.2 Mechanical parameters of Flax TRMs tested in tension

Specimen	P_{max} (N)	$disp_{max}$ (mm)	σ_{max} (MPa)	ε_{max} (%)	$\sigma_{textile}$ (MPa)	$\sigma_{max}/f_{textile}$ (%)	Failure mode
Flax TRM-1L-T-1	1426	15.7	4.75	5.2	233	81	A
Flax TRM-1L-T-2	1161	16.3	3.87	5.4	190	66	B
Flax TRM-1L-T-3	1249	14.5	4.16	4.8	204	71	B
Flax TRM-1L-T-4	1181	15.7	3.94	5.2	193	67	B
Mean	1255	15.6	4.18	5.2	205	72	–
Co.V. (%)	10	5	10	5	10	10	–
Flax TRM-2L-T-1	2189	16.8	4.56	5.6	179	62	A
Flax TRM-2L-T-2	2591	17.9	5.40	6.0	212	74	A
Flax TRM-2L-T-3	2615	16.8	5.45	5.6	214	75	B
Flax TRM-2L-T-4	2443	16.5	5.09	5.5	200	70	B
Flax TRM-2L-T-5	2312	14.7	4.82	4.9	189	66	A
Mean	2430	16.5	5.06	5.5	199	69	–
Co.V. (%)	7	7	7	7	7	7	–

Table 4.3 Stress and strain corresponding to the different stage transition points

Specimen	σ_1 (MPa)	ε_1 (%)	σ_2 (MPa)	ε_2 (%)	$\sigma_{1,textile}$ (MPa)	$\sigma_{2,textile}$ (MPa)
Flax TRM-1L-T-1	–	–	2.47	2.9	–	121
Flax TRM-1L-T-2	–	–	2.22	3.3	–	109
Flax TRM-1L-T-3	–	–	2.13	2.7	–	104
Flax TRM-1L-T-4	1.65	0.11	2.92	4.3	81	143
Mean	1.65	0.11	2.43	3.3	81	119
Co.V. (%)	–	–	15	22	–	15
Flax TRM-2L-T-1	1.56	0.08	3.00	3.9	61	118
Flax TRM-2L-T-2	1.62	0.09	2.66	3.6	64	104
Flax TRM-2L-T-3	1.37	0.08	2.99	3.6	54	117
Flax TRM-2L-T-4	1.76	0.11	2.90	3.6	69	114
Flax TRM-2L-T-5	1.50	0.07	2.50	2.9	59	98
Mean	1.56	0.08	2.81	3.5	61	110
Co.V. (%)	9	19	8	10	9	8

emphasises two aspects: it confirms the maximum load to be strongly affected by the amount of textile; it points out that by increasing the fibre volumetric reinforcement ratio interlaminar shear phenomena were not recorded, hence any decrease in textile-to-mortar bond properties related to higher concentration of stress within the cross section occurred. In terms of maximum displacement, and strain, the two series exhibited a quite similar trend.

The values of stress and strain in correspondence to the two transition points between the various stages were evaluated as well and are reported in Table 4.3.

The mean value of the stress corresponding to the occurrence of the first crack resulted lower to the flexural strength of the mortar (3.58 MPa). Such aspect is attributed to the different loading conditions between the TRM tensile test and the mortar characterisation flexural test, and to possible geometric or test set-up imperfection that may anticipate the development of the mortar cracking. With respect to the transition point between stages II and III, it was observed a similar value with respect to the two series in terms of strain, while a slightly higher value was observed in terms of stress in the Flax TRM-2L-T series with respect to the other one. Such result is distorted by the cross section adopted, i.e. the entire TRM area. By referring to the dry textile cross section the trend would invert, with slightly lower values recoded in the Flax TRM-2L-T series.

The maximum stress achieved in the textile, in both the series, was about 70% of the maximum tensile strength of the flax textile. No differences, between the two series, in the exploitation ratio value, confirmed that the mechanical properties remained unchanged by increasing the amount of textile. Although the tensile strength of the textile was not attained, the failure mode was characterised by a sequence of sudden drops of the load, attributed to the threads' rupture (Fig. 4.11c) (mode B). This discrepancy may be attributed to the different loading conditions of

Fig. 4.11 **a** Failure mode of specimens Flax TRM-1L-T-4 and **b** Flax TRM-2L-T-4, **c** close-up of fabric threads rupture

the textile when tested alone with respect to the case in which it is restrained by the TRM mortar. Moreover, during the test the slippage between mortar and textile may damage the fibres. Lastly, as shown in Chap. 3, the inclusion of flax within the lime-based mortar causes a slight reduction of its strength.

4.1.4 Analysis of the Crack Pattern

In the first stage the specimens are characterised by an elastic behaviour, and a roughly uniform state of stress along their length can be assumed. The onset of the cracking process occurs with the mortar failure in tension in the weakest section of the specimen, whose position is uncertain. The next cracks position depends on this initial random event. In correspondence to a crack, the entire load is borne by the textile. Such load is transferred from the textile to the mortar by means of shear stress at the fibre-to-matrix interface. According to this behaviour, the stress within the mortar increases with the distance from a cracked section, up to a value next to the mortar tensile strength, so as, to cause to cracking of the mortar section. This process of transfer of normal stress from the textile to the matrix, mainly depends on the fibre-to-matrix bond properties, and on the quantity of interfacial surface available (i.e. the volumetric fibre reinforcement ratio). In fact, by looking at Fig. 4.11a, b it is clear that the presence of a larger number of cracks in the specimen characterised by a higher amount of textile.

Therefore, the analysis of the crack pattern, and of the distance between cracks and their opening, may represent an important tool to better understand the mechanical behaviour of the TRM and to get information about the interaction between the composite components. Such investigation was carried out by means of the Digital Image Correlation (DIC) analysis. By means of an image processing software [17], the displacement field along a longitudinal section in the middle of the external surface of the composite was obtained for each specimen, and with an interval of 20s during the entire test. Such analysis was carried out on a portion of the specimen, 200 mm long, placed in the middle of the free length. By selecting significant images, corresponding to the transition between two different stages, or to the occurrence of cracks, it is possible to get curves describing the evolution of the displacements along the specimen and during the test. In Figs. 4.12 and 4.13 such curves are shown with respect of two representative specimens of the series Flax TRM-1L-T and Flax TRM-2L-T (curves from all the specimens are shown in Appendix). The displacement field is represented by the use of different colours. Each curve represents the image that corresponds to the strain reported on the right side. It can be observed that when the crack occurs, a relative displacement between the two sides of the interested section is recorded. As a result, each step of the curve corresponds to the occurrence of a crack. In the final configuration, the represented by the upper curve in the graph, the distance between the cracks can be easily evaluated being the distance between two consecutives steps. Each step represents the width of a crack. By getting such curve at different values of the strain, it is possible to get the order in which the cracks developed, and the evolution of each crack opening at different strain levels. The evolution of the cracks opening during the test is plotted in Figs. 4.14 and 4.15 with respect of two representative specimens of the series Flax TRM-1L-T and Flax TRM-2L-T. The curves derive from the displacement field along a gauge length of

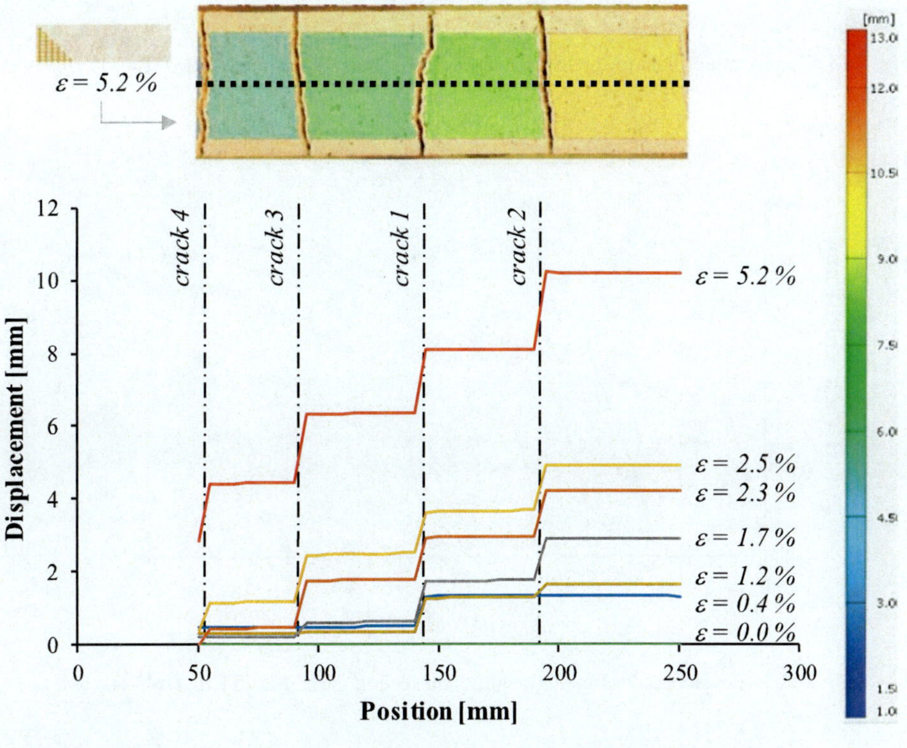

Fig. 4.12 Displacement field of the mean longitudinal section (spec. Flax TRM-1L-T-1)

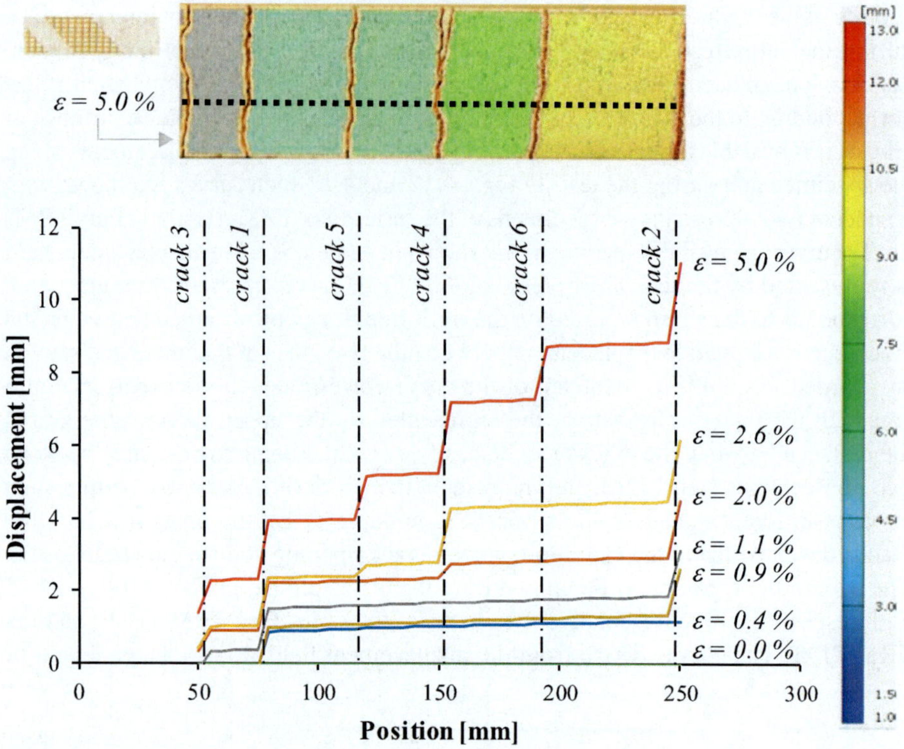

Fig. 4.13 Displacement field of the mean longitudinal section (spec. Flax TRM-2L-T-4)

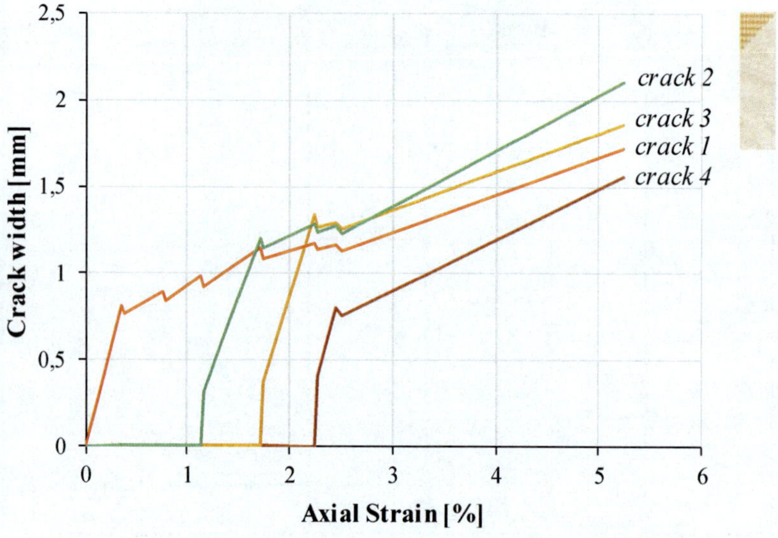

Fig. 4.14 Cracks development of mean longitudinal section (spec. Flax TRM-1L-T-1)

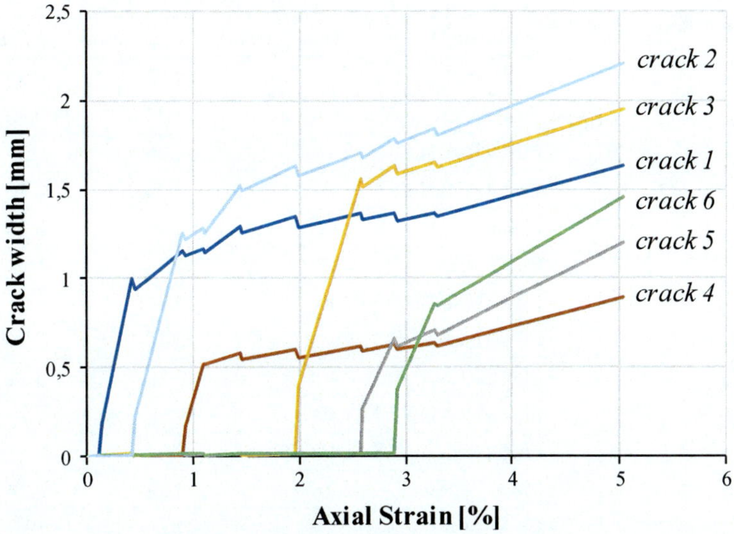

Fig. 4.15 Cracks development of mean longitudinal section (spec. Flax TRM-2L-T-4)

200 mm in the middle of the sample. During the tests some cracks occurred out of this length. Normally the strain interval between the development of two consecutive cracks is quite constant for all of them. However, by looking at Figs. 4.14 and 4.15 such a thing is not verified for all the cracks. This aspect is due to the development of cracks out of the gauge length.

The analysis of crack pattern, hence the mean crack spacing between the cracks, is not affected by the neglecting of the cracks developed out of the gauge length. By analysing the graphs, for each specimen the mean value of the distance between two consecutives cracks, the mean crack spacing $\delta_{0,mean}$, and the maximum value of the crack opening, w_{max}, were extracted and reported in Table 4.4. By looking at the mean values of the maximum crack width it is highlighted that both reinforcement configurations adopted, by respectively using one and two plies of flax textiles, led to a similar maximum crack opening in correspondence to the specimen failure. Such evidence was expected being the stress within the textile not significantly different between the two series of specimens.

With respect to the mean value of the distance between two cracks, δ_{0mean}, it was observed that the mean value concerning the Flax TRM-1L-T series was linearly proportional to the analogous distance observed in the Flax TRM-2L-T series, with a proportionality coefficient factor equal to the ratio between the two volumetric reinforcement ratios, vf_2/vf_1 (being vf_1 and vf_2 respectively equal to 2.04 and 2.55%). Such dependence of the distance between cracks from the fibre volumetric reinforcement ratio, emphasises that the two series of specimens were characterised by the same fibre-to-matrix bond behaviour.

A method to analyse the development of the cracks of TRM in tensile tests, consists in tracing the evolution of the mean distance between the cracks during the

Table 4.4 Analysis of cracks main parameters

Specimen	$\delta_{0,mean}$ (mm)	w_{max} (mm)	τ (MPa)
Flax TRM-1L-T-1	47	2.03	0.23
Flax TRM-1L-T-2	55	2.33	0.18
Flax TRM-1L-T-3	44	2.26	0.26
Flax TRM-1L-T-4	57	2.70	0.26
Mean	51	2.33	0.23
Co.V. (%)	12	12	16
Flax TRM-2L-T-1	40	2.22	0.21
Flax TRM-2L-T-2	31	2.48	0.29
Flax TRM-2L-T-3	40	2.80	0.21
Flax TRM-2L-T-4	39	2.19	0.29
Flax TRM-2L-T-5	45	2.07	0.21
Mean	39	2.35	0.24
Co.V. (%)	13	12	19

test [16]. Figure 4.16 shows such curve with respect to a representative specimen of the series Flax TRM-2L-T, evaluated as a logarithmic interpolation of discrete data. The graph shows that with the occurrence of cracks during the test, the mean distance between them decreases, until the end of the test at which the value corresponds to the mean distance δ_{0mean}. In Fig. 4.17 the development of the mean crack spacing is plotted for each specimen for both series. As expected, the comparison shows that the specimens of the series Flax TRM-2L-T, in the final part, resulted to be below the curves concerning the other series. Such trend pointed out that, taken a generic

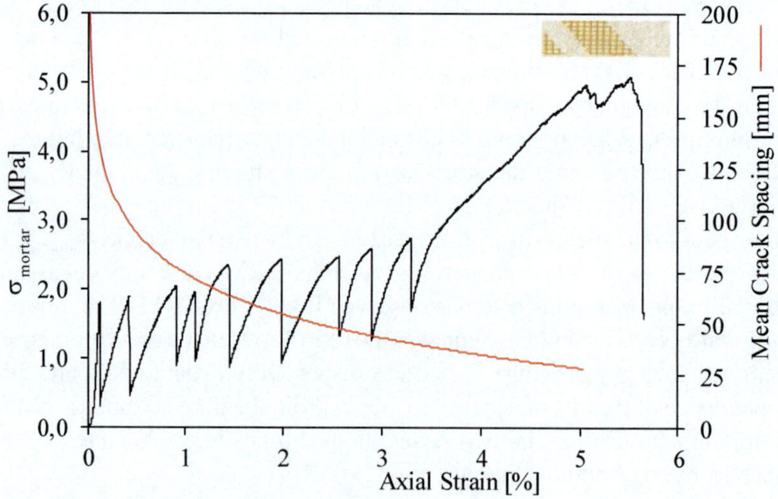

Fig. 4.16 Mean crack spacing as a function of the strain (spec. Flax TRM-2L-T-4)

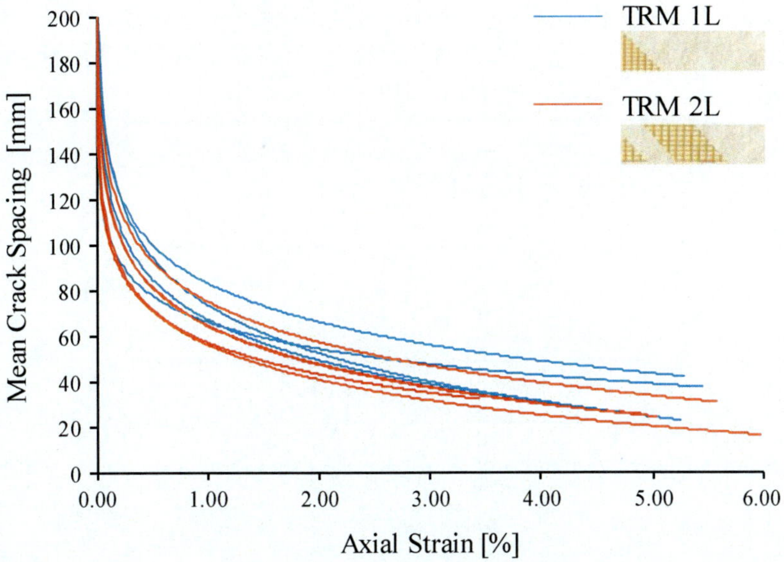

Fig. 4.17 Mean crack spacing as function of strain (series Flax TRM-1L-T and 2L-T)

value of the strain, the specimens reinforced by a higher amount of textile exhibited a higher number of cracks. This aspect is of fundamental importance in the assessment of the efficiency of the composite: the higher is the number of cracks developed, the more the system responds as a composite material in which stress and strain are distributed along the entire coupon, and the drops of load observed in the second stage reduce their size.

By analysing the crack pattern at the end of the test for each specimen, it is possible to get information about the shear stress at the fibre-to-matrix interface. An approximated estimation model is proposed as follows.

In the elastic phase it can be assumed that both the matrix and the textile contribute in bearing the applied force, F (Eq. 4.1):

$$F = \sigma_m(z)A_m + \sigma_t(z)A_t \tag{4.1}$$

where σ_m and σ_t represent the matrix and textile normal stresses, A_m and A_t are the matrix and the textile cross-sections. By assuming null the contribution of the mortar in a cracked section, and by assuming a constant distribution of the shear stresses transferred from the textile to the mortar, the stresses distribution in the damaged portion in proximity to a crack can be approximated with the graphs in Fig. 4.18.

Moving along the z, the portion of load transferred from the fibres to the mortar is equal to the integral of the shear stress τ along the z (Eq. 4.2):

$$\sigma_t(z)A_t = F - \tau \cdot p \cdot z \tag{4.2}$$

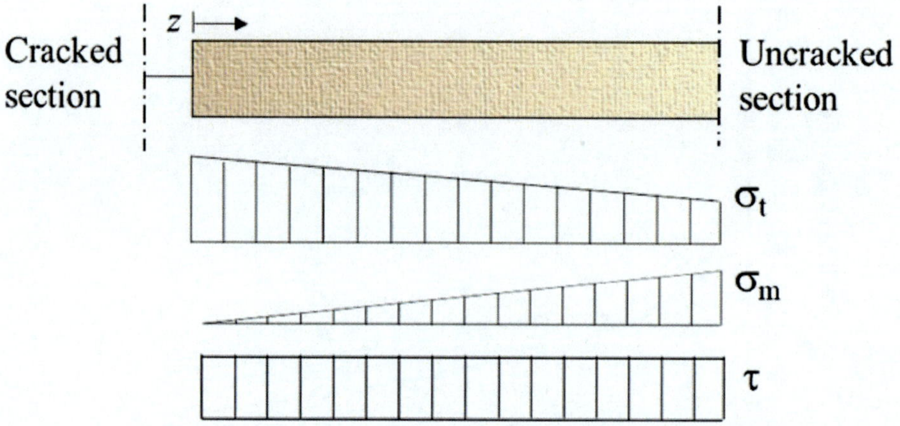

Fig. 4.18 Approximated distribution of normal and shear stresses in the proximity of a cracked section

where p is the lateral surface per unit length of the fibres (assessed in Chap. 3).

The length δ_0 can be defined asn the value of z corresponding to the achievement of the maximum tensile strength in the mortar ($\sigma_m = f_{t,m}$, being $f_{t,m}$ the uniaxial tensile strength of the mortar).

By substituting Eq. 4.1 in Eq. 4.2. the following relationship between the stress τ and the distance δ_0 is obtained (Eq. 4.3):

$$\tau = f_{t,m} \cdot A_m/(p \cdot \delta_0) \tag{4.3}$$

From the experimental evidence, a value can be assigned to $\delta_{0,}$ as the smaller distance between two consecutive cracks (assessed by means of DIC analysis). By adopting Eq. 4.3. an estimation of the shear stress at the fibre to matrix interface, τ, can be calculated. The values of such estimation are shown in Table 4.4. According to the results obtained, the shear stress assumed almost the same value for the two series of specimens.

It is worth pointing out that the shear values assessed represent an approximated estimation, due the hypothesis considered with respect to the interfacial stress distributions. In fact, in the in the Stages II and III there is a strain incompatibility between matrix and fibres, due to the relative displacements and damaging of the interface surface. Moreover, in this configuration the shear stress distribution is far to be constant over the z direction. However, the same order of magnitude of the shear stress with respect to the two series of specimen confirms that no significant changes in terms of bond properties occurred in the TRM system when the amount of textile was increased.

4.2 Shear Bond Test on Flax TRM-to-Masonry Systems

4.2.1 Materials and Methods

Shear bond (SB) tests were carried out by adopting the single-lap shear test configuration aiming to reproduce an intermediate crack-induced failure mode, characterised only by shear stresses distributed on the interfacial surfaces of the system components. In order to be consistent with studies present in the literature, the recommendations of the Committee TC 250 CSM RILEM [9] were respected in the definition of the experimental setup and the mechanical parameters of the study.

The specimens consisted of prismatic masonry substrates comprising five clay bricks and four 10 mm-thick mortar joints, externally strengthened by Flax TRM strips. In line with the tensile characterisation, two series of specimens were considered (Table 4.5):

- Flax TRM-1L-SB: specimens characterised by five bricks masonry element externally strengthened by the composite in the reinforcement configuration Flax TRM-1L, subjected to shear bond test (Fig. 4.19);
- Flax TRM-2L-SB: specimens characterised by five bricks masonry element externally strengthened by the composite in the reinforcement configuration Flax TRM-2L, subjected to shear bond test (Fig. 4.19).

Two of the specimens of the series Flax TRM-1L-SB were discarded because of an incorrect recording of the displacements during the test. That is why there is an inconsistency in the number of specimens of each series.

The Flax TRM strips, with a bond length of 260 mm, were placed centrally along the vertical axis of the prism and at a distance of 20 mm from the upper side of the support. The textile free length at the loaded edge of the specimen was of 300 mm. The reinforcement was applied on the masonry elements by adopting the same manufacturing process utilised for tensile test specimens.

The clay bricks, belonging to the commercial line "Rosso Vivo A6R55 W" produced by San Marco-Terreal, Valenza AL, Italy [8], were 250 mm long, 120 mm wide, and 55 mm thick, with a compression strength (perpendicular to the larger face) of 18 MPa (manufacturer technical specifications).

Table 4.5 Main properties of the shear bond test series of specimens

Series	Number of specimens	TRM section (mm × mm)	A_{mortar} (mm^2)	$A_{textile}$ (mm^2)	vf (%)	f_{textil} (MPa)
Flax TRM-1L-SB	3	5 × 60	300	6.1	2.04	286
Flax TRM-2L-SB	5	8 × 60	480	12.2	2.55	

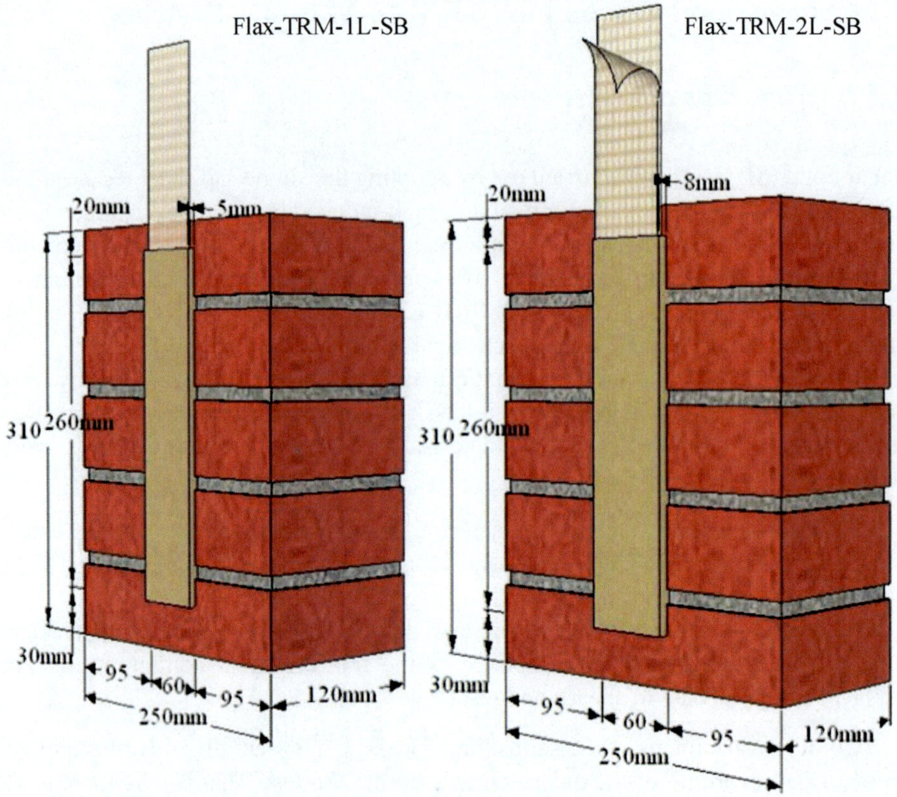

Fig. 4.19 Flax TRM-to-masonry specimens subjected to shear bond test: Flax-TRM-1L-SB, characterised by one layer of flax textile; Flax TRM-2L-SB, with two layers

A low strength hydraulic lime-based mortar was adopted to implement the masonry bed joints. This mortar was designed in consistency with the mortar traditionally adopted in the implementation of masonry structures. The volumetric composition of the mortar was chosen according to the Italian Guidelines [20] recommendations to reproduce a hydraulic lime mortar with low strength: 2 parts of binder, 4 parts of sand, and 1.3 parts of water.

The main mechanical properties, assessed according to standard specifications [11], consisted in a compression and flexural strength respectively equal to 2.45 MPa (Co.V. 14%) and 0.50 MPa (Co.V. 37%).

The flax textile mechanical properties were shown in Chap. 3. The hydraulic lime-based mortar adopted to perform the Flax TRM systems was characterised by a compression and a flexural strength respectively equal to 10.55 MPa (Co.V. 8%) and 3.56 MPa (Co.V. 5%), assessed after 28 days of curing from the casting.

The experimental investigation was performed at the Laboratory of Composite Materials for Constructions (LMC2) of the University Claude Bernard Lyon 1. Single-lap shear tests were carried out by using a Zwick/Roell 1475 electromechanical testing machine, with a 50 kN capacity in tension and compression,

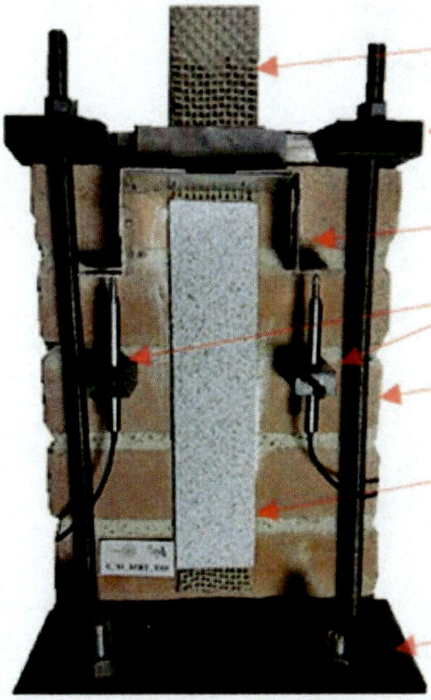

Fig. 4.20 Shear bond test set-up

in displacement control using a cross-head rate of 0.3 mm/min. A steel system, consisting of two metallic plates connected by four threaded rods, locked the masonry prisms so that no rotation of the reinforced substrate could occur (Fig. 4.20). Steel plates bonded to the end of the flax textile were placed by a clamping steel device. To avoid load eccentricity and, consequently, out-of-plane tensile stresses (peeling) at the interface between the masonry substrate and the reinforcing system, each specimen was duly placed to ensure that the textile was aligned with the load direction, with no intentional eccentricity. Moreover, a hinge was placed between the clamping steel device and the head of the universal machine so that no bending moments could be transmitted to the FlaxTRM system.

In order to record possible slipping of the textile within the mortar matrix, an aluminium device, named "Ω" for its shape, was bonded to the flax textile at a distance of 5 mm from the TRM edge. Two displacement transducers (Linear Variable Differential Transformer, LVDTs) bonded to the masonry substrate, were used to monitor any motion of the "Ω" plate during the test.

4.2.2 Results and Discussion

The shear bond test response is expressed in terms of stress-slip curves (Figs. 4.21 and 4.22). The normal stress, σ, was obtained by dividing the applied load for the cross-section of textile; the slipping, s, of the textile within the mortar matrix was computed as the average value among the two displacements recorded by the transducers, LVDTs, monitoring the "Ω" device.

The main parameters obtained from the tests are shown in Table 4.6. P_{max} represents the maximum load, s_{max} represents the corresponding main displacement of

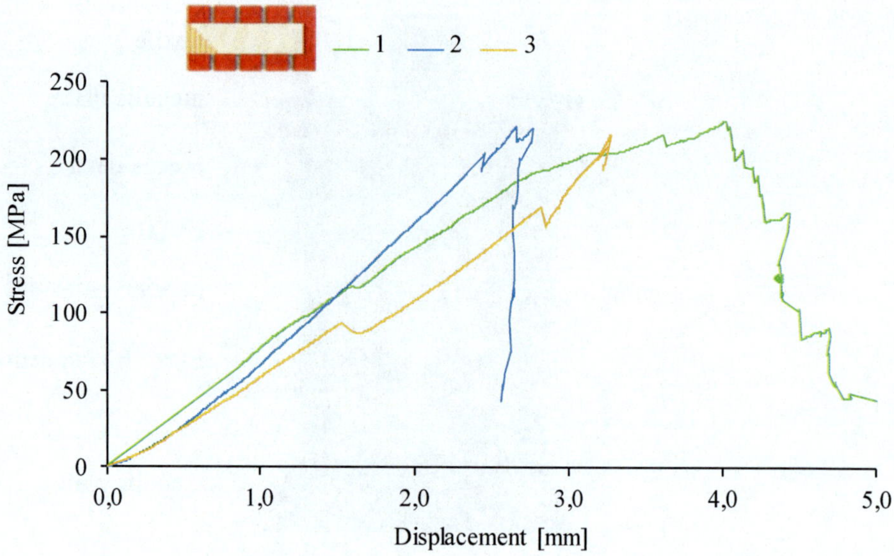

Fig. 4.21 Stress-slip curves of shear bond tests concerning the Flax TRM-1L-SB series

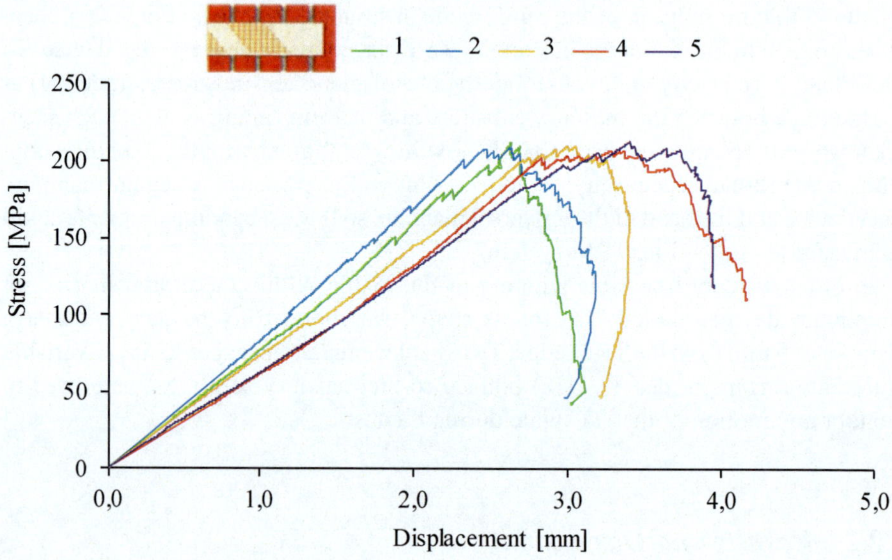

Fig. 4.22 Stress-slip curves of shear bond tests concerning the Flax TRM-2L-SB series

the "Ω" device, σ_{max} is the corresponding normal main stress within the textile. In order to quantify the fibres exploitation, the ratio between the maximum stress, σ_{max}, and the tensile strength of the textile, $f_{textile}$, is reported for each specimen.

The stress-displacement curves show that both the series of specimens exhibit the same trend characterised by a linear elastic branch up to the achievement of the

Table 4.6 Mechanical parameters of Flax TRM-to-masonry systems tested in shear

Specimen	P_{max} (N)	s_{max} (mm)	σ_{max} (MPa)	$\sigma_{max}/f_{textile}$ (%)	Failure mode
Flax TRM-1L-SB-1	1379	4.0	226	79	E
Flax TRM-1L-SB-2	1355	2.7	222	77	E
Flax TRM-1L-SB-3	1325	3.3	217	76	E
Mean	1353	3.3	221	77	–
Co.V. (%)	2	21	2	2	–
Flax TRM-2L-SB-1	2598	2.6	212	74	E
Flax TRM-2L-SB-2	2551	2.6	209	73	E
Flax TRM-2L-SB-3	2572	3.1	210	73	E
Flax TRM-2L-SB-4	2530	3.4	207	72	E
Flax TRM-2L-SB-5	2604	3.4	213	74	E
Mean	2571	3.0	210	73	–
Co.V. (%)	1	13	1	1	–

maximum capacity. Such response may be interpreted as a combination of the linear elastic deformation of the textile portion placed between the "Ω" device and the TRM system, and possible relative displacements between mortar and fibres within the composite. The post peak phase is mainly characterised by a sequence of drops of the load due to the rupture of the single flax threads until the failure of the specimen.

For all specimens the failure mode consisted in the rupture of the flax threads in the free length of the textile (mode E) (Fig. 4.23a). Such rupture did not involve a specific cross section: it was rather due to a process of redistribution of the stress within the textile after each thread rupture; at each step, the most charged thread randomly failed at the weakest point of its length. Figure 4.23b shows a close-up in which

a) 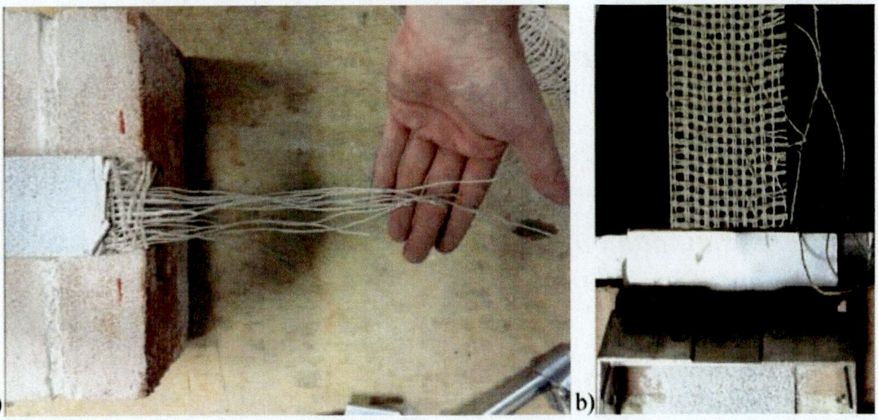 b)

Fig. 4.23 **a** Failure mode of a representative specimen (Flax TRM-2L-SB-5); **b** Close-up of the flax threads' failure in the free length few threads already failed while the others were stretched due to a concentration of stress.

Both series of specimens achieved an exploitation of the fibres of about 75%. Although the fibres achieved tensile failure, the mean maximum stress never reached the value recorded during the dry textile tensile tests. Such behaviour, already showed also in the case of tensile tests, may be attributed to two aspects: on one side the free length is of 300 mm, while the dry textile was tested with a gauge length of 150 mm, as also shown in Chap. 2 the higher is the size of the specimen, the higher is the probability of having geometric and mechanic imperfection, the lower is the tensile capacity; on the other side, while in dry textile tensile tests the fabric is perfectly clamped at both the edges, in the case of shear bond test one of the sides is embedded in the mortar, allowing for deformation of the textile also within the TRM system, hence causing a lower strength.

It is worth to notice that the average tensile capacity observed in shear bond tests is slightly higher of that exhibited by composite specimens tested in tension, even if the same geometry of the Flax TRM system was adopted. Such experimental evidence may be explained by the fact that in tensile tests the distribution of the stress within the composite during the test create a damage at the matrix to fibre interface with a deterioration of the threads that result in a lower global capacity of the system.

The shear bond test showed that, among the different capacities associated to the possible failure mechanisms, i.e. loss of bond strength at the TRM-to-substrate interface, loss of bond at the textile to mortar interface, loss of strength of the textile in tension, the one associated to the latter is the lower. On one hand, this aspect emphasises that such failure mode is the most brittle, hence it has to be properly considered on the analysis of the global behaviour of structural elements, that are often required to exhibit a ductile behaviour [15]. On the other hand, such mode of failure allows to increase the load capacity of the system by increasing the textile amount since the good adhesion at the fibre-to-matrix interface does not lead to a premature failure of the composite system due to the slippage of the textile within the mortar matrix. The latter, in fact, results as one of the main limitations showed by most of high strength synthetic fibres (such as carbon and basalt), and in such cases, the improvement of the textile-to-mortar bond, by means of fibre treatments or matrix modifications, results to be a challenging task [14].

4.2.3 Determination of the Flax TRM Designing Conventional Strength

As also discussed in paragraph 2.1.4, according to Italian guidelines for designing of rehabilitation interventions by means of TRMs [6], a conventional strength, to be used for designing and verification problems, can be derived by the mechanical characterisation strength. Specifically, such strength consists of a value of the normal stress, $\sigma_{lim,conv}$, corresponding to the maximum strength deriving from shear bond

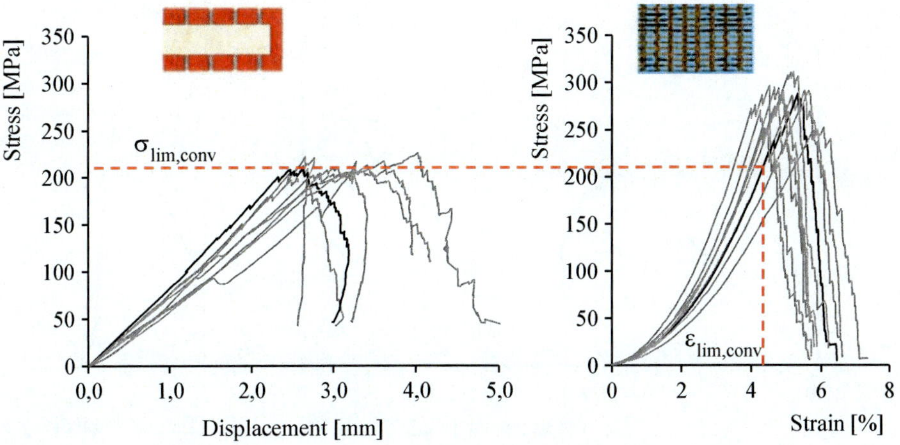

Fig. 4.24 Determination of Flax TRM conventional strength according to the instructions CNR-DT 215, 2018

tests, and in a value of the axial strain, $\varepsilon_{lim,conv}$, corresponding to the value of the strain corresponding to the stress $\sigma_{lim,conv}$ in the stress-strain tensile response of the dry textile.

By considering both series of specimens the conventional maximum stress, $\sigma_{lim,conv}$, results equal to 215 MPa. The conventional strain is assessed as the average among the values corresponding to all the stress-strain curves of the series Flax-strip-6 cm analysed in Chap. 3 (Fig. 4.24).

The conventional strain, $\varepsilon_{lim,conv}$, assumes the mean value of 4.58%. However, within the instructions it is assumed that the tensile textile response is linear. In the case of plant fibres, a less stiff branch is typically exhibited at low strain levels, due to the morphology of the fibres at a microscale, and to the arrangement of the textile on a macroscale.

Being this stiffness, E_f, determined in the linear branch, it is necessary to define a modified value of the limit strain, ε^*, as the difference between strain values ε_1 and ε_2 (Fig. 4.25), so that the product between the conventional limit strain and the mean stiffness results to be equal to the conventional limit stress, $\sigma_{conv,lim}$.

The graphical assessment of the conventional strain is shown in Fig. 4.25 where, to make the image clearer, two representative specimens are plotted: a shear bond test response characterised by a maximum stress of about the $\sigma_{lim,conv}$, and a flax textile tensile response characterised by a linear stiffness of about E_f.

The outcoming mean values are: conventional limit stress, $\sigma_{lim,conv}$, of 215 MPa; a mean dry stiffness of the textile, E_f, of 9.03 GPa (Chap. 3); and a modified conventional limit strain, ε^*, equal to 2.4%.

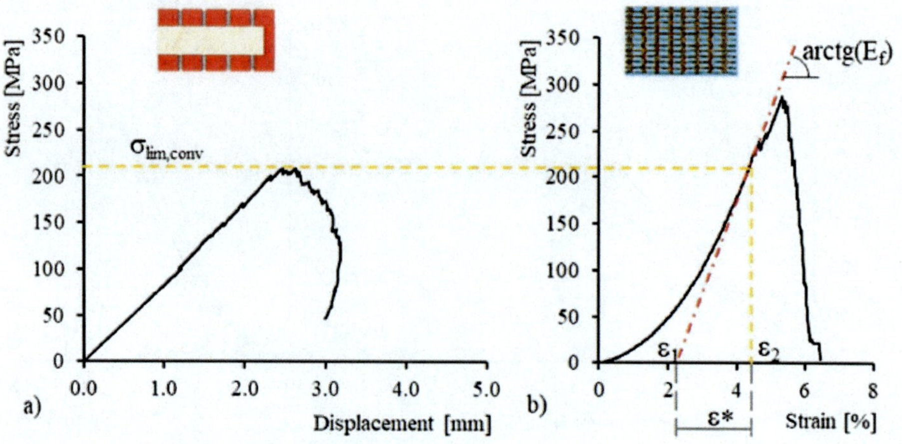

Fig. 4.25 Modified value of the conventional limit strain assessed by comparing shear bond test and flax tensile test representative responses

4.3 Durability

The durability of plant fibre-based TRMs represents one of the main concerns that feeds scepticism in the use of the composite system. Such distrust seems to be due to a lack of knowledge rather than to exhaustive scientific outcomes. This is the reason why, with the aim of providing a comprehensive analysis of the effectiveness of Flax TRMs, such aspect was addressed within the study. However, being mechanical aspects the core of the discussion, durability issues were boarded with a secondary attention, with the aim of at least proving Flax TRMs to comply with the existing standards and guidelines. As a consequence of the ageing analysis conducted on flax threads, discussed in Chap. 3, in this paragraph the attention is focused directly on Flax TRM coupons.

4.3.1 Materials and Methods

The qualification procedure adopted consisted in the exposure of Flax TRM specimens in specific ageing conditions chosen according to the acceptance criteria [1]. The durability performance was computed by comparing the tensile behaviour of the specimens tested after the period of exposure, with the response of reference specimens stored in ambient conditions. The composite material components (namely, mortar and textile) were subjected to the same treatments to quantify their influence in the global response.

The following ageing protocols were adopted:

- Ambient: the specimens were stored in ambient conditions for 3000 h (125 days) and were used as reference;

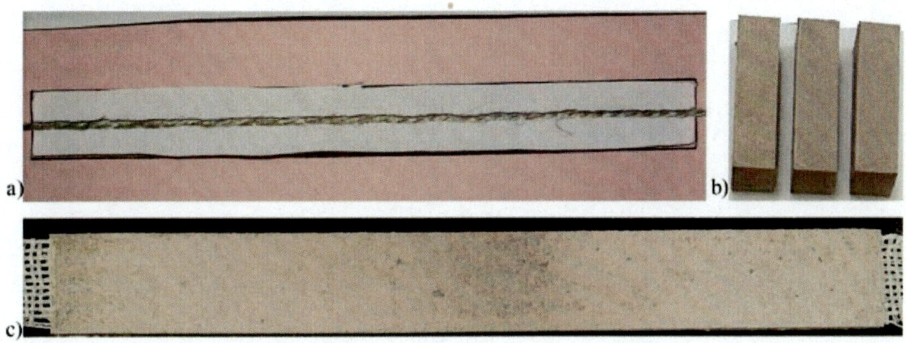

Fig. 4.26 Samples for durability analysis: **a** Flax-T; **b** H-Lime-Matrix; **c** Flax TRM-1L-T

- Water: the specimens were immersed in deionised water for a period of 3000 h (125 days);
- Salt-water: specimens were immersed for 3000 h (125 days) in a saline aqueous solution including 3.5% in weight of sodium chloride (NaCl) to reproduce the world's ocean seawater average salinity;
- Alkaline solution: the specimens were immersed for a period of 3000 h (125 days) in an alkaline environment with 9.5 PH level, reproduced by adding sodium hydroxide (NaOH) to the deionised water.

For each conditioning environment, in line with procedures of the literature [19], the following specimens were considered:

- Flax-T: consisting in flax threads tested in tension with a gauge length of 100 mm (the same type of specimen adopted in Chap. 2) (Fig. 4.26a);
- H-Lime-Matrix: consisting in mortar prisms (40 × 40 × 160 mm^3) to be tested in bending and compression in accordance to the EN 196-1 [11] (Fig. 4.26b);
- Flax TRM-1L-T: consisting of TRM coupons reinforced by the configuration Flax TRM-1L, characterised by 5 mm of thickness, 60 mm of width and 500 mm of length (of which 300 mm of gauge length), and tested in tension (Fig. 4.26c).

The main characteristics of all series of specimens, curing conditions and the respective type of test are listed in Table 4.7.

Tensile tests on fibres and bending tests on mortar, were performed by means of a Universal Testing Machine by using respectively a 1 kN load cell, with a displacement rate of 4 mm/min (Fig. 4.27a), and a 5 kN load cell, with a displacement rate of 0.2 mm/min (Fig. 4.27b).

The immersion of the Flax TRM specimens in the different solutions was performed by taking care of avoiding the direct absorption of water through flax textile. To this purpose, silicon was applied before the immersion in the water solution to seal the two edges of the specimens (Fig. 4.28). Tensile tests on the flax threads were carried out at the Department of Industrial Engineering of University of Salerno. Mortar compression tests (Fig. 4.27c) and TRM tensile tests (Fig. 4.27d) were carried out by means of a Zwick Roell Schenck Hydropolus S56, with maximum capacity

Table 4.7 Series of specimen considered in the experimental durability analysis

Series	Conditioning environment	Conditioning period	No of spec.	Geometry	Type of test	Textile section
Flax T-Amb	Ambient	3000 h	14	150 mm	Tensile test	0.25 mm^2
Flax T-W	Water	3000 h	14			
Flax T-S	Salt	3000 h	14			
Flax T-Alk	Alkaline	3000 h	14			
H-Lime-Matrix-Amb	Ambient	–	3/6	40 × 40 × 160 mm^3/ ~40 × 40 × 40 mm^3	Three-point bending test/Compression test	–
H-Lime-Matrix-W	Water	3000 h	3/6			
H-Lime-Matrix-S	Salt	3000 h	3/6			
H-Lime-Matrix-Alk	Alkaline	3000 h	3/6			
Flax TRM-1L-T-Amb	Ambient	–	5	5 × 60 × 500 mm^3	Tensile test	6.12 mm^2
Flax TRM-1L-T-W	Water	3000 h	5			
Flax TRM-1L-T-S	Salt	3000 h	5			
Flax TRM-1L-T-Alk	Alkaline	3000 h	5			

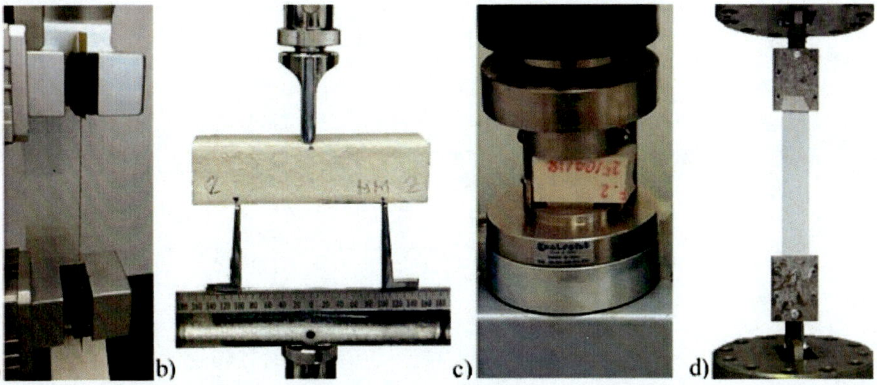

Fig. 4.27 Test set-up of series: **a** Flax T; **b** H-Lime-Mortar (bending); **c** H-Lime-mortar (compression); **d** Flax TRM-1L-T

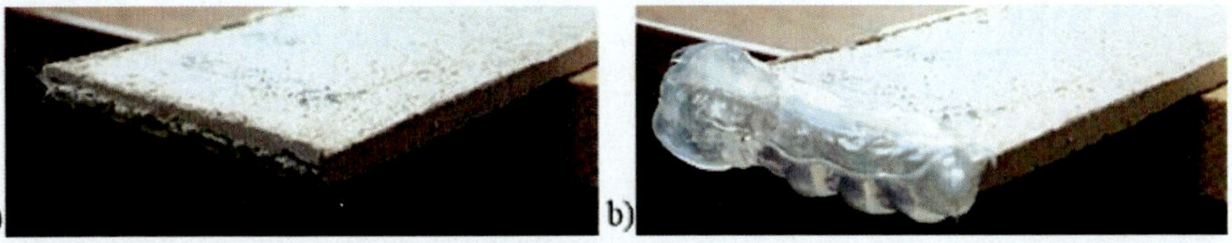

Fig. 4.28 **a** Flax TRM-1L-T edge before sealing; **b** Sealing of the edge

of 630 kN, respectively with a displacement rate of 0.3 mm/min and 0.25 mm/min, at the Strength Laboratory of University of Salerno.

4.3.2 Results and Discussion

The effect of application of the ageing protocols on the flax threads is analysed in terms of tensile strength, $\sigma_{t,max}$, of the corresponding strain, ε_{max}, and of stiffness, E (Table 4.8) (Coeff. of variation in round brackets). In terms of strength a slight increase of the mean maximum stress was observed in the series Flax-T-W. The series Flax-t-S and Flax-T-Alk exhibited a slightly less stiff response. The effect of different environmental exposures on the mortar is analysed in terms of compression, $\sigma_{c,max}$, and flexural strength, $\sigma_{f,max,}$ of the prisms (Table 4.9).

The specimens aged in water and in alkaline solution exhibited a compression strength slightly higher of that shown by the reference series. For both TRM components it can be affirmed that no significant changes were brought by the application of the durability protocols, and few variations of the mechanical properties may be attributed to the statistical variability of the results. The TRMs response is shown

Table 4.8 Parameters concerning the Flax-T specimens

Series	$\sigma_{t,max}$ (MPa)	ε_{max} (%)	E (GPa)
Flax T-Amb	283.67	3.3	9.07
	[13]	[14]	[14]
Flax T-W	309.38	3.5	9.13
	[14]	[15]	[7]
Flax T-S	294.02	3.6	7.98
	[14]	[7]	[9]
Flax T-Alk	267.74	3.7	7.09
	[9]	[8]	[11]

Table 4.9 Parameters concerning the H-Lime-Mortar specimens

Series	$\sigma_{f,max}$ (MPa)	$\sigma_{c,max}$ (MPa)
H-Lime-Mortar-Amb	4.81	11.79
	[3]	[4]
H-Lime-Mortar-W	4.05	13.73
	[5]	[5]
H-Lime-Mortar-S	4.54	12.21
	[7]	[7]
H-Lime-Mortar-Alk	4.39	14.10
	[16]	[7]

in terms of stress-strain curves for each series of specimens: the reference specimens in ambient conditions (Fig. 4.29), the specimens aged in water (Fig. 4.30), the specimens immersed in saltwater (Fig. 4.31) and the specimens exposed to alkaline environment (Fig. 4.32).

As expected, the response was characterised by the tri-phase behaviour widely discussed in paragraph 0. The main mechanical parameters concerning Flax TRM tensile tests for the different ageing environments, are shown in (Tables 4.10 and 4.11). In line with the mechanical response exhibited by the TRM components, no significant changes were observed also in the behaviour of the whole composite tested in tension.

More in details, the three environmental exposures conferred to the composite a slight increase of maximum strength. In fact, the textile exploitation ratio (assessed with respect to the non-aged textile), of about 70% in the reference series, gets values included in the range between the 80% and 85% in the aged series. It is worth highlighting that the exploitation ratio of the series Flax TRM-1L-T-Amb, tested after 124 days, was the same of the one observed in the series Flax TRM-1L-T (paragraph 4.1), tested after 28 days.

Such aspect confirms that no significant degradation of the textile in contact with the hydraulic lime mortar occur in the period between 28 and 124 days.

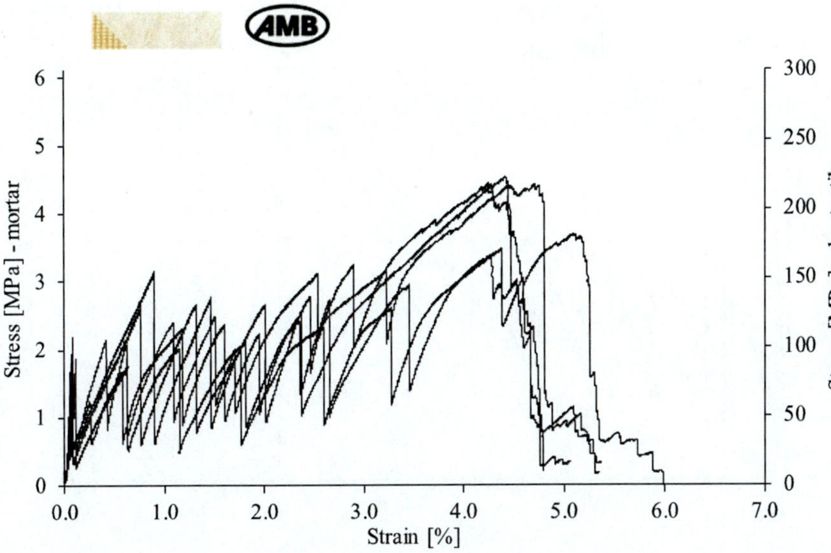

Fig. 4.29 Stress-strain curves of the series Flax TRM-1L-T-Amb

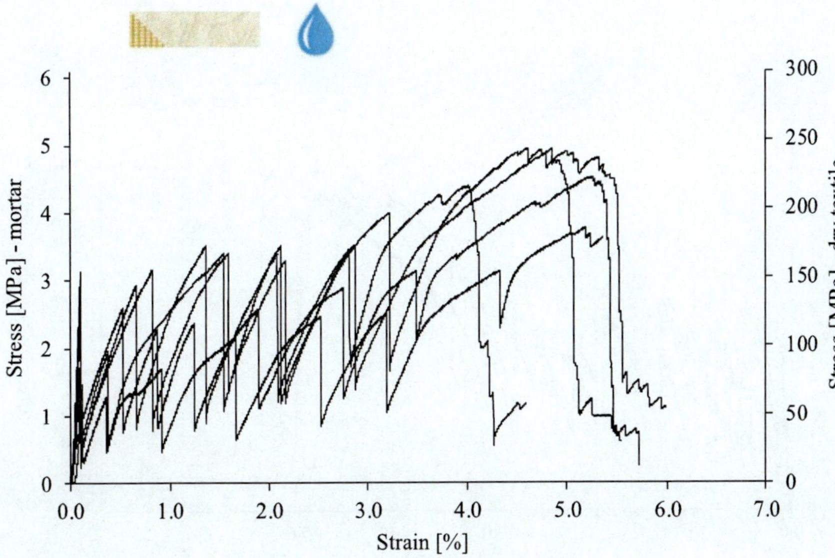

Fig. 4.30 Stress-strain curves of the series Flax TRM-1L-T-W

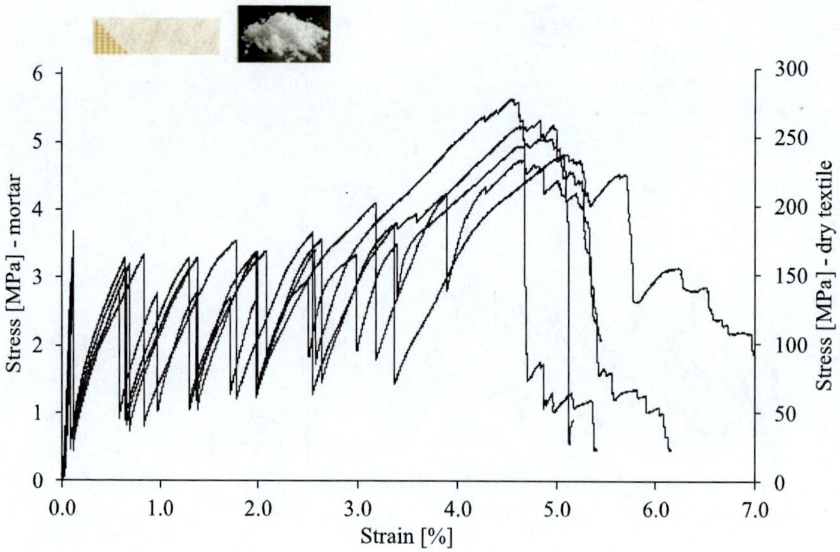

Fig. 4.31 Stress-strain curves of the series Flax TRM-1L-T-S

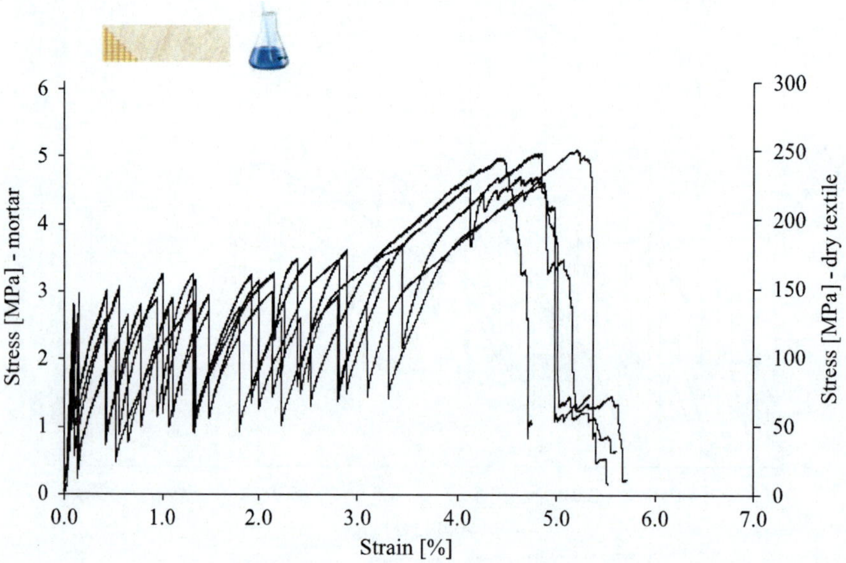

Fig. 4.32 Stress-strain curves of the series Flax TRM-1L-T-Alk

Table 4.10 Parameters concerning the Flax TRM-1L-T specimens

Series	P_{max} (N)	$disp_{max}$ (mm)	σ_{max} (MPa)	ε_{max} (%)	$\sigma_{textile}$ (MPa)	$\sigma_{max}/f_{textile}$ (%)
Flax TRM-1L-T-Amb	1224	13.6	4.08	4.5	200	70
	(15)	(8)	(15)	(8)	(15)	(15)
Flax TRM-1L-T-W	1380	15.0	4.60	5.0	226	79
	(7)	(16)	(7)	(16)	(7)	(7)
Flax TRM-1L-T-S	1540	14.4	5.13	4.8	252	88
	(7)	(4)	(7)	(4)	(7)	(7)
Flax TRM-1L-T-Alk	1468	14.4	4.89	4.8	240	84
	(5)	(6)	(5)	(6)	(5)	(5)

Table 4.11 Stress and strain corresponding to the different stage transition points

Series	σ_1 (MPa)	ε_1 (%)	σ_2 (MPa)	ε_2 (%)	τ (MPa)
Flax TRM-1L-T-Amb	1.85	0.1	2.87	3.4	0.24
	(13)	(23)	(14)	(11)	(18)
Flax TRM-1L-T-W	2.13	0.1	3.29	3.6	0.17
	(34)	(11)	(16)	(6)	(18)
Flax TRM-1L-T-S	3.12	0.1	3.73	3.6	0.21
	(11)	(14)	(8)	(15)	(32)
Flax TRM-1L-T-Alk	2.58	0.1	3.51	3.6	0.23
	(16)	(29)	(9)	(9)	(18)

The low increase in the mortar strength led to transition points between the stage I and II, and between the stages II and III, characterised by higher stress, σ_1 and σ_2. Such experimental evidence may be attributed to the fact that the conditioning in water facilitates the hydration processes of the mortar.

The specimens exhibited a failure mode characterised by several mortar cracks, causing significant drops of the load, until the tensile failure of the textile in proximity on one of the fissures (Fig. 4.33) (modes A and B). In all the specimens the mean distance between consecutive cracks assume a similar value. Being the textile reinforced ratio constant in all the series, it can be assumed that the ageing processes did not significantly affect the fibre-to-matrix interface bond behaviour. In Table 4.11 the mean fibre-to-matrix interfacial shear stress, τ, assessed with the simplified model proposed in paragraph 4.1.4, is reported for each series of specimens. It can be seen that no significant differences were observed.

As a matter of fact, since no significant decrease of strength were observed after the exposure of the Flax TRM samples to the different ageing protocol provided by the AC 434-1011-R1 [1], it can be affirmed that the composite system complies with the acceptance criteria generally considered for the qualification of the mechanical behaviour in the long terms.

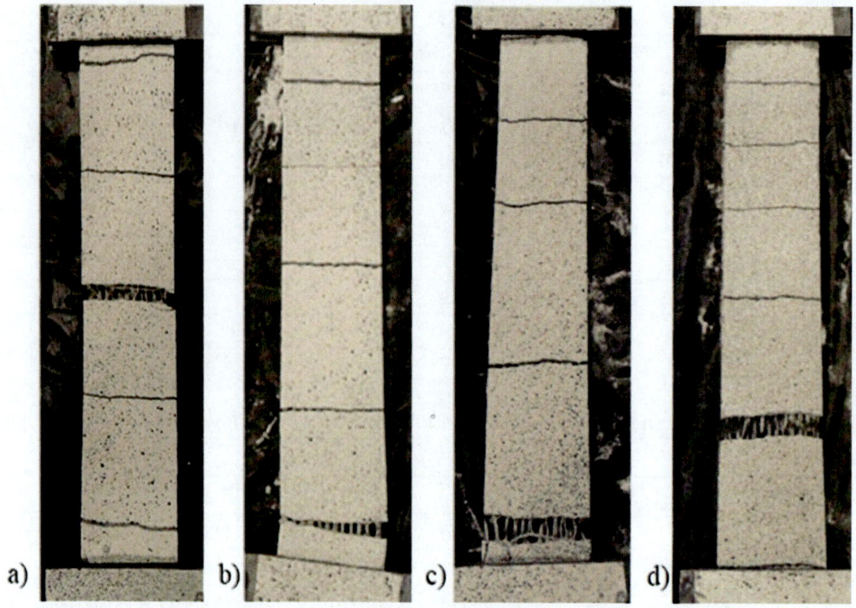

Fig. 4.33 Failure mode of representative specimens of the series: **a** Flax TRM-1L-T-Amb; **b** Flax TRM-1L-T-W; **c** Flax TRM-1L-T-S; **d** Flax TRM-1L-T-Alk

4.4 Conclusions

The present chapter aims at providing an exhaustive mechanical characterisation of the Flax TRM system under investigation, at the scale of analysis of the composite. Both tensile behaviour and the shear bond performance with a masonry substrate were considered. Moreover, the characterisation was enriched with a preliminary experimental analysis of the durability of Flax TRM composites. The main findings of the experimental analysis are summarised as follows:

- Tensile tests carried out on Flax TRM coupons showed that the response in tension is characterised by the typical three stages. However, unlike what is observed in high strength fibre based TRMs, the stress-strain response of Flax TRMs exhibited significant drops of the stress in correspondence to crack occurrence, due to the low initial stiffness of plant fibre textiles, to bond properties at the fibre-to-matrix interface and to the utilised mortar mix design;
- All specimens tested in tension exhibited a failure mode characterised by rupture of the fibres, with an exploitation ratio of about 70%, due to the different load conditions of the textile when tested alone with respect to the case in which it is restrained by the mortar;
- By varying the fibre volume fraction no reduction in strength was observed, showing that the textile grid was well designed allowing the mortar to correctly

penetrate through the voids. Such statement was confirmed by analysing the fibre-to-matrix shear stress assessed by means of a simplified model;
- Digital Image Correlation analysis confirmed to be a valuable and powerful technique to investigate in depth the crack pattern developed on the surfaces of TRM composites tested in tension;
- With respect to the Flax TRM-to-masonry substrate shear bond behaviour, the experimental evidence showed an elastic response of the textile in the free length of the loaded edge ending up in a rupture of the textile in such area. This behaviour, occurred in both reinforcement configurations, emphasising that the system is characterised by a good adhesion at the fibre-to-mortar interface with respect to the tensile strength of the flax textile itself, and that a bond length equal to 260 mm is adequate for preventing TRM-to-masonry debonding;
- The experimental investigation carried out to analyse the durability of Flax TRM composites showed that the system, exposed to a conventional ageing protocol, complies with the acceptance criteria proposed by standards in the field.

The promising outcomes of the comprehensive investigation carried out at the composite scale of analysis, encourage the employment of Flax TRM systems as reinforcement of structural masonry elements. Experimental studies with the application of such composite system on structural elements, are needed in order to confirm the potential of the strengthening technique, and perhaps to highlight the limitations of the system. The latter is a fundamental aspect: the experimental study, together with the encouraging results, pointed out that there is still room for improvements of the mechanical performance of plant based TRMs. The definition of the crucial aspects on which it is necessary to focus the attention plays a key role in the process of improvement and refinement of the reinforcing technique.

References

1. AC 434-1011-R1 (2018) Acceptance criteria for masonry and concrete strengthening using fiber-reinforced cementitious matrix (FRCM) composite systems. Birmingham, AL, ICC Evaluation Service
2. ACI 549.4R-13 (2013) Guide to design and construction of externally bonded fabric-reinforced cementitious matrix (FRCM) systems for repair and strengthening concrete and masonry structures. ACI Committee 549, Farmington Hills, MI
3. ASTM C136/C136M-14 (2014) Standard test method for sieve analysis of fine and coarse aggregates. West Conshohocken, PA, USA, ASTM International
4. Bilotta A, Ceroni F, Lignola GP, Prota A (2017) Use of DIC technique for investigating the behaviour of FRCM materials for strengthening masonry elements. Compos B Eng 129:251–270
5. Cevallos OA, Olivito RS (2015) Effects of fabric parameters on the tensile behaviour of sustainable cementitious composites. Compos B Eng 69:256–266
6. CNR-DT 215/2018 (2018) Istruzioni per la Progettazione, l'Esecuzione ed il Controllo di Interventi di Consolidamento Statico mediante l'utilizzo di Compositi Fibrorinforzati a matrice inorganica, *Consiglio Nazionale delle Ricerche – Commissione per lo studio per la predisposizione e l'analisi di norme tecniche relative alle costruzioni* (Italy)

7. Company Website (2019) Innovations.srl. Available http://www.innovationssrl.it. Accessed 29 Dec 2019
8. Company Website, San Marco Terreal (2020). Available https://www.sanmarco.it/. Accessed 3 Jan 2020
9. de Felice G, Aiello MA, Caggegi C, Ceroni F, De Santis S, Garbin E, Gates N, Hojdys Ł, Krajewski P, Kwiecień A, Leone M, Lignola GP, Marzetti C, Oliveira D, Papanicolaou C, Poggi C, Triantafillou T, Valluzzi MR, Viskovic A (2018) Recommendation of RILEM technical committee 250-CSM: test method for textile reinforced mortar to substrate bond characterization. Mater Struct 51(4):95
10. EN 1015-3: 1999 (1999) Methods of test mortar for Masonry. Determination of consistence of fresh mortar (by flow table). Brussels, Belgium, European Committee for Standardization
11. EN 196-1:1994 (1994) Methods of testing cement—part 1: determination of strength. European Committee for Standardization, Brussels, Belgium
12. Ferrara G, Martinelli E (2018) Tensile behaviour of textile reinforced mortar composite systems with flax fibres. In: Proceedings of the 12th fib international Ph.D. symposium in civil engineering, 863–869
13. Ferrara G, Coppola B, Di Maio L, Martinelli E (2018) Physical and mechanical characterization of natural fibres and fabrics as reinforcement for composite systems. In: 9th international conference on fibre-reinforced polymer (FRP) composites in civil engineering, CICE 2018, 250–257
14. Ferrara G, Caggegi C, Gabor A, Martinelli E (2019) Experimental study on the adhesion of basalt textile reinforced mortars (TRM) to clay brick masonry: the influence of textile density. Fibers 7(12):103
15. Ferrara G, Caggegi C, Martinelli E, Gabor A (2020) Shear capacity of masonry walls externally strengthened using Flax-TRM composite systems: experimental tests and comparative assessment. Constr Build Mater 261:120490
16. Fidelis MEA, de Andrade Silva F, Toledo Filho RD (2014) The influence of fibre treatment on the mechanical behaviour of jute textile reinforced concrete. Key Eng Mater 600:469–474
17. GOM Correlate (2017) Digital image correlation Software. https://www.gom.com/3d-software/gom-correlate.html. Accessed 31 Dec 2019
18. Kouris LAS, Triantafillou TC (2018) State-of-the-art on strengthening of masonry structures with textile reinforced mortar (TRM). Constr Build Mater 188:1221–1233
19. Nobili A (2016) Durability assessment of impregnated glass fabric reinforced cementitious matrix (GFRCM) composites in the alkaline and saline environments. Constr Build Mater 105:465–471
20. NTC 2018 (2018) Norme Tecniche per le Costruzioni. Ministry of the Infrastructures and Transports of the Italian Republic
21. Olivito RS, Cevallos OA, Carrozzini A (2014) Development of durable cementitious composites using sisal and flax fabrics for reinforcement of masonry structures. Mater Des 57:258–268
22. Saidi M, Gabor A (2020) Adaptation of the strain measurement in textile reinforced cementitious matrix composites by distributed optical fibre and 2D digital image correlation. Strain 56:12335
23. Technical Sheet (2019) FIDCALX NHL5®, Betoncino Strutturale di Calce Idraulica Naturale per L'impregnazione di Tessuti per il Rinforzo Strutturale. Available http://www.fidiaglobalservice.com. Accessed 29 Dec 2019

Chapter 5
Mechanical Behaviour of Masonry Walls Externally Strengthened by Flax TRM Composites

As shown in the previous chapters, the use of plant fibres as reinforcement in mortar-based composites represents a promising solution to improve the sustainability of the material. However, due to the specific mechanical properties of plant textiles, inferior in regard to "conventional" ones, (i.e. lower strength and stiffness) some differences in the mechanical behaviour emerged with respect to TRM composite performed by conventional textiles. Therefore, further studies are needed to better understand the behaviour of these types of TRM as a possible technological solution for enhancing shear capacity of masonry walls.

To respond to this need, this chapter reports experimental observations obtained on clay-brick masonry walls externally strengthened by the Flax TRM systems under study. Specifically, the reinforced masonry structural elements are subjected to diagonal compression tests in order to assess the contribution of the intervention to the shear capacity of the system. After discussing the experimental evidence into detail, a comparison of results with similar studies present in the literature, and with models proposed by standards in this area, is shown as well.

5.1 Methods

With the aim of assessing the diagonal shear (or tensile) capacity of masonry elements externally strengthened by Flax-TRM composites, diagonal compression tests were carried out. The walls, loaded along one diagonal, exhibited a tension failure with the specimens splitting apart perpendicular to the loading direction. Tests were conducted at the Laboratory of Composite Materials for Construction (LMC2), University Claude Bernard Lyon 1.

The analysis was carried out considering three series of specimens (Fig. 5.1):

- USW, unstrengthened walls providing the reference behaviour;
- SW-1L, masonry walls externally strengthened by the composite system Flax-TRM-1L;

G. Ferrara, *Flax-TRM Composite Systems for Strengthening of Masonry*, Springer Theses,
https://doi.org/10.1007/978-3-030-70273-1_5

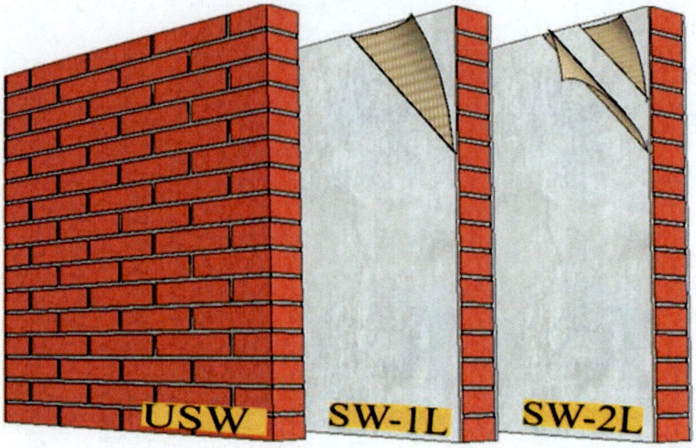

Fig. 5.1 Diagonal Compression Test specimens: UnStrengthened (USW); Flax TRM-1L Strengthened (SW-1L) and Flax TRM-2L Strengthened (SW-2L) walls

- SW-2L, masonry walls externally strengthened by the composite system Flax-TRM-2L.

The TRM reinforcement system and the masonry substrate were performed with the same characteristics of the specimens subjected to tensile and shear bond test, whose experimental outcomes are discussed in Chap. 3. The bricks were arranged in the running bond pattern, a monolayer configuration characterised by a single row of bricks in thickness of one head. The low strength hydraulic lime-based mortar adopted for wall bed and head joints was characterised by a flexural strength of 0.94 MPa (Co.V. 20%) and by a compression strength of 4.11 MPa (Co.V. 21%). The main mechanical properties of the hydraulic lime mortar utilised as matrix of the Flax TRM systems consisted in a flexural strength of 3.13 MPa (Co.V. 13%) and a compression strength of 11.13 MPa (Co.V. 8%).

The composite, applied on the walls of the series USW-1L and USW-2L, was uniformly spread on the two surfaces of the masonry elements. The composite application procedure comprised the follow steps: a thin layer of mortar was applied on the masonry surface (Fig. 5.2a); the textile was embedded in the mortar laying on the wall surface, by guaranteeing a proper tightening of the fabric (Fig. 5.2b); the next layer of mortar was applied covering the flax textile; possible application of further textile and mortar layers for the reinforcing multi-ply configuration; smoothing of the outer layer of mortar (Fig. 5.2c).

The main characteristics of the three series of specimen are listed in Table 5.1. The table also includes the reinforcement rate, defined as the maximum tensile capacity of the textile included in a TRM strip, 1 m wide, considering all the threads exhibiting their maximum strength. The diagonal compression test set-up is shown in Fig. 5.3.

Tests were carried out by using a machine, with maximum load capacity in compression of 500 kN, provided with a load cell placed on the top of the specimen

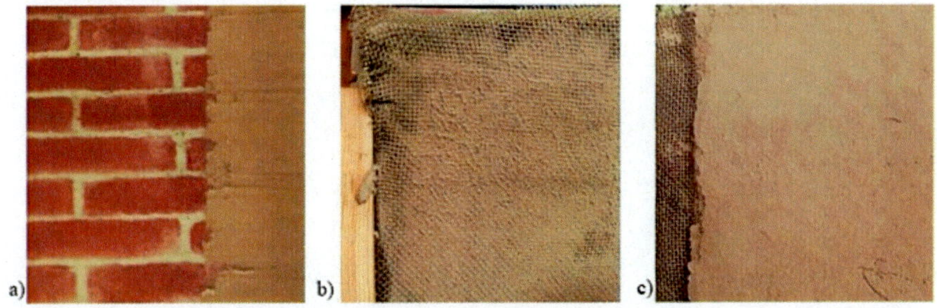

Fig. 5.2 Implementation of the composite: **a** application of the first layer of mortar; **b** impregnation of the flax textile in the mortar; **c** application of the outer layer of mortar and smoothing of the surface

Table 5.1 Proprieties of the series for diagonal compression tests

Series	No. spec.	Masonry type	Wall size t × w × h (mm^3)	Reinforc. type	Type of mortar	Thick (mm)	Reinf. rate (kN/m)
USW	3	Clay brick	120 × 1030x1030	–	–	–	–
SW-1L	3	Clay brick	120 × 1030x1030	Flax TRM-1L	Hydraulic lime	5	31
SW-2L	3	Clay brick	120 × 1030x1030	Flax TRM-2L	Hydraulic lime	8	62

recording the load values. Diagonal compression tests were controlled in displacement with an advancement rate of 1 mm/min. A steel loading shoe was placed at each corner of the vertical diagonal, to transmit the charge from the machine to the specimen. This metallic device was properly designed to avoid significant concentration of stresses in the loading area, hence, to avoid local failure phenomenon. To this purpose, a steel device having a thickness of 30 mm (high stiffness) and a length of bearing of 150 mm was adopted. The full connection between the wall and the steel shoe was guaranteed by the application of a thin layer of high strength fast-setting cement-based mortar over the entire length of bearing. The displacements along the diagonal directions, shortening of the vertical one and lengthening of the horizontal one, were recorded via two linear displacement transducers placed on the wall surface with a gauge length of 1000 mm.

One of the two sides of the walls was treated in order to get, during the test, images to be used for a Digital Image Correlation (DIC) analysis. The treatment consisted in a white background on which black dots were stochastically sprinkled by means of a spray can. The camera was placed at a focal distance of 260 cm. The acquisition rate was of 2 ph/s and the resolution of the images was of 2560 × 2048 pixels (0.51 mm/pix). The analysed surface size was of 1 m × 1 m (Fig. 5.4).

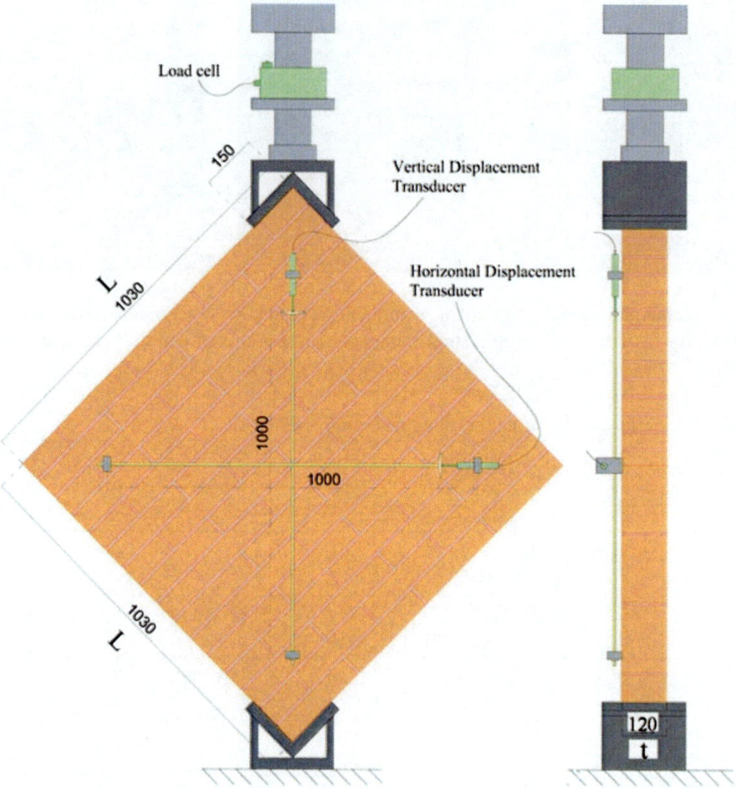

Fig. 5.3 Diagonal compression Test set-up

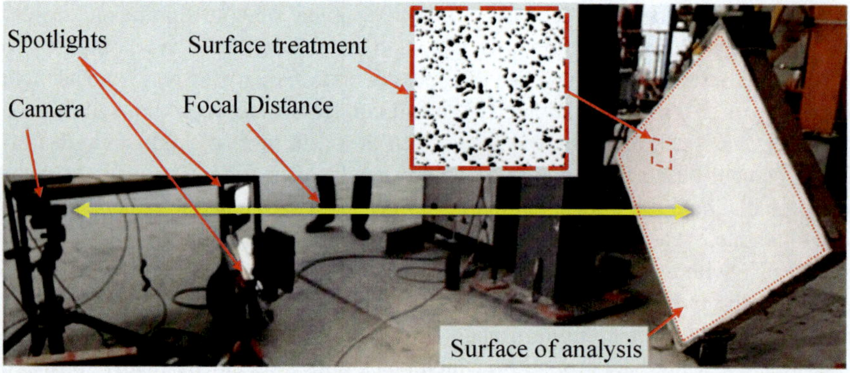

Fig. 5.4 Digital image correlation equipment for walls tested in diagonal compression

5.2 Results and Discussion

5.2.1 Failure Mode

Diagonal compression loading conditions cause a concentration of stresses along the tensile principal direction (the vertical loading diagonal). As consequence, the failure of the walls involves the development of several vertical cracks in the middle of the panel. The reference behaviour of the unstrengthened wall included a brittle failure mode. Quasi-vertical cracks, either following the mortar bed joints or breaking the clay bricks, caused the splitting in two parts of the wall (Fig. 5.5a). The crack pattern developed on the walls externally strengthened by the Flax-TRM composite started from the middle of the panel, showing a similar trend with respect to the reference samples (Fig. 5.5b). Specimens of the series SW-1L and SW-2L exhibited the failure in tension of the flax threads placed in the transversal cracked sections (Fig. 5.6). This phenomenon, occurring in the post-peak phase, emphasises an overall mechanical response of the wall complying the experimental evidence from the test on the local scale on analysis. Both in tensile and shear bon tests, the tensile capacity of the flax textile mainly governed the ultimate failure of the composite system (Chap. 4).

This failure mode, on one hand confirms the low tensile capacity of flax fibres (for example with respect to the common high strength fibres), on the other hand it highlights the effective bond between the system components. Both TRM-to-masonry and fibres-to-mortar bond capacity are higher with respect to the tensile strength of the textile. In addition, it was shown that the amount of textile did not affect this failure mechanism. Therefore, it seems possible to even improve the global shear capacity of the system by increasing the textile amount. A different isolated behaviour was shown by the specimen SW-1L-1, characterised by a local debonding of the composite consisting in the detachment of the Flax-TRM from the substrate in

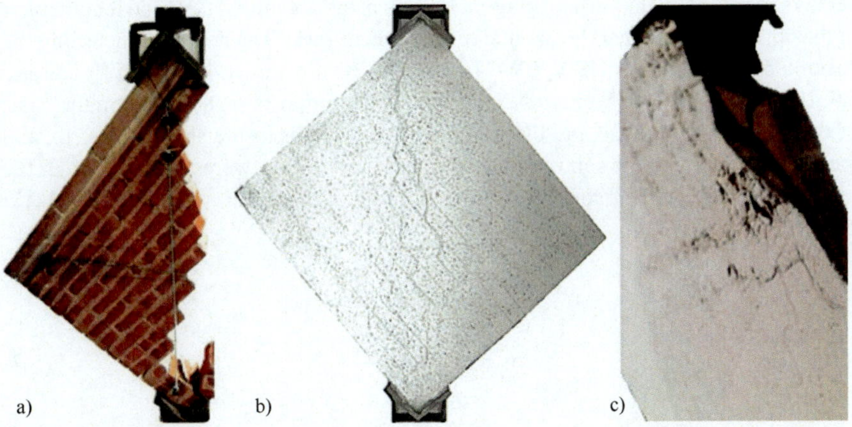

Fig. 5.5 Failure mode of samples: **a** USW-2; **b** SW-1L-3; **c** SW-1L-1 (local debonding)

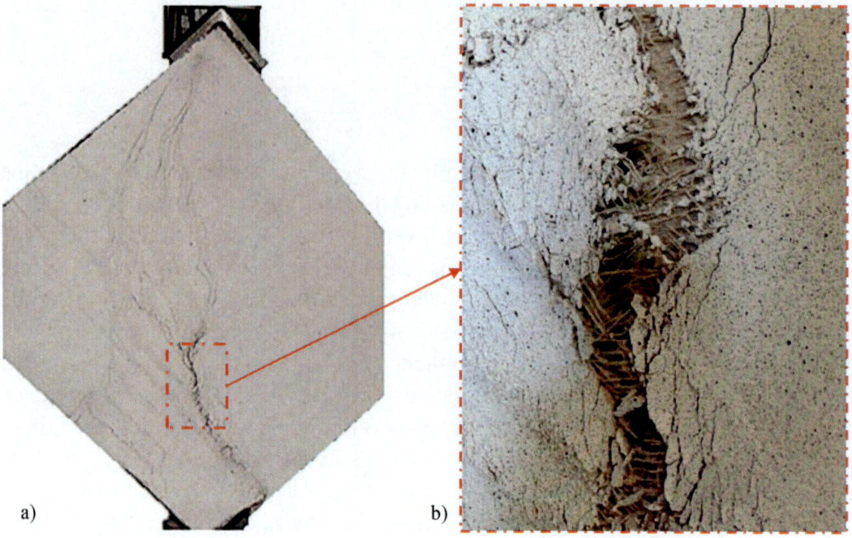

Fig. 5.6 **a** Failure of the specimen SW-2L-3; **b** close-up of the fibres rupture

proximity of one of the edges of the wall (Fig. 5.5c). The local buckling of the reinforcing layer may be due to a lower composite-to-substrate bond due to inaccuracy during the application of the Flax-TRM system on the wall.

5.2.2 Shear Stress Versus Shear Strain Response

Diagonal compression tests results are expressed in terms of curves applied load versus shortening of the vertical diagonal (vertical displacement ΔV) and lengthening of the horizontal diagonal (horizontal displacement ΔH). The results, concerning the specimens of the series USW, SW-1L and SW-2L are shown in Figs. 5.7, 5.8 and 5.9. In accordance with standards ruling diagonal compression tests [1], on the basis of the recorded data, load and displacements, the values of the shear stress, τ_0, and the shear strain, γ, in the centre of the masonry assemblage were assessed, under the hypothesis of pure shear conditions in the middle of the panel, via Eqs. 5.1 and 5.2:

$$\tau_{0,ASTM} = \frac{P}{\sqrt{2} \cdot A_n} \tag{5.1}$$

$$\gamma = \frac{\Delta V + \Delta H}{L_g} \tag{5.2}$$

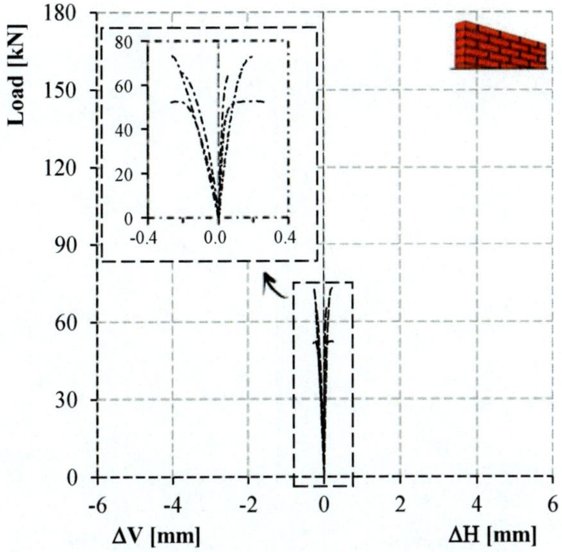

Fig. 5.7 Test Load-Vertical/Horizontal displacements curves, series USW

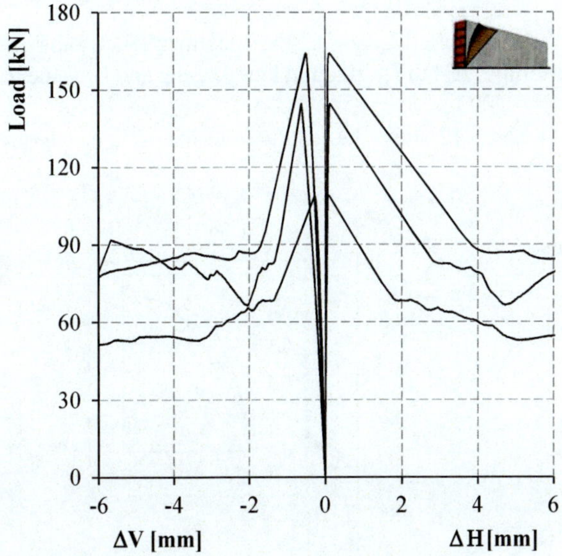

Fig. 5.8 Load-Vertical/Horizontal displacements curves, series SW-1L

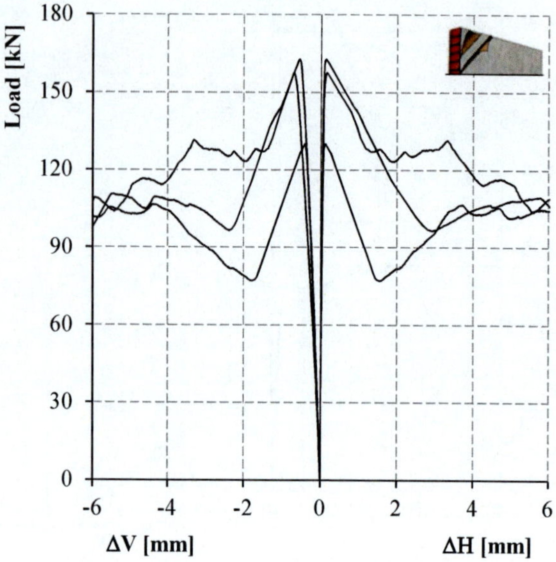

Fig. 5.9 Load-Vertical/Horizontal displacements curves, series SW-2L

in which P represents the applied Load, A_n the specimen's net transversal section, ΔV the vertical shortening, ΔH the horizontal lengthening and L_g is the gauge length of 1 m.

Figures 5.10, 5.11 and 5.12 show the results in terms of τ_0, shear stress, and, γ,

Fig. 5.10 Shear stress—shear strain curves for series USW

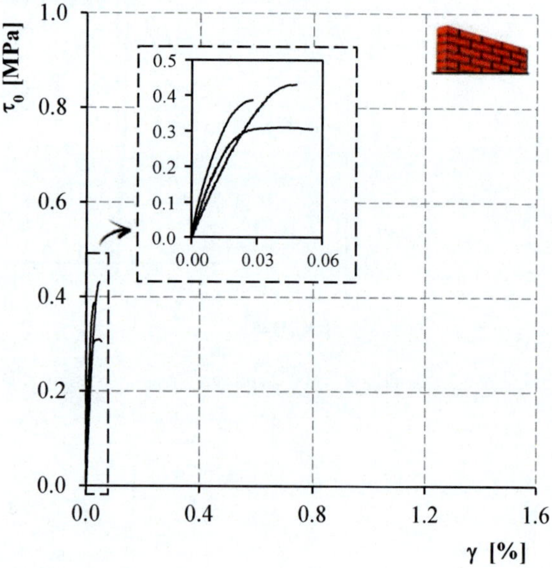

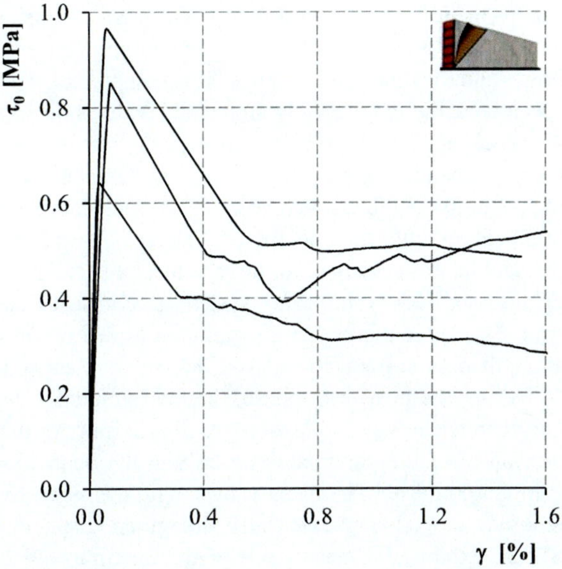

Fig. 5.11 Shear stress—shear strain curves for series SW-1L

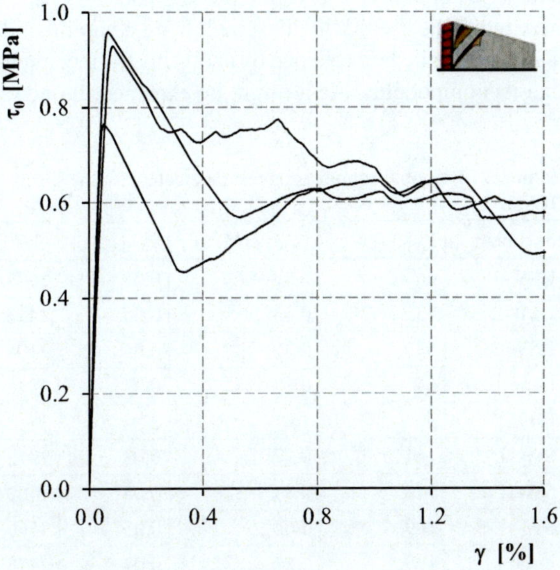

Fig. 5.12 Shear stress—shear strain curves for series SW-2L

shear strain, respectively for the series USW, SW-1L and SW-2L.

The curves concerning the unstrengthened walls, USW specimens, showed a linear response followed by a short plastic branch. The specimens failed in a brittle manner due to development of a marked crack along the loaded diagonal (Fig. 5.10). The SW-1L and SW-2L specimens showed a pseudo-elasto-plastic response until the achievement of the maximum load. The cracks, causing the end of the elastic phase, develop along the vertical diagonal and through the entire cross section including the bricks and the mortar of the Flax TRM composite system. Unlike the unstrengthened specimens, the externally strengthened walls showed a post peak phase, consisting in the capacity of the system to bear significant loads after the elastic phase, in correspondence to high values of the displacement, due to the tightening of the textile in proximity of the cracked sections. A considerable loss of strength was observed immediately after the opening of the cracks in the middle of the panel. However, a residual load is sustained for large values of shear strains due to the beneficial effect of the Flax TRM reinforcing system, that ensures the post-failure integrity of the wall. The latter represents a significant aspect in view of implementation of these strengthening systems to improve the structure capacity towards horizontal actions.

During the post peak phase, the load gradually decreased with the progressive failure in tension of the flax threads in proximity of the cracked zone. Table 5.2 lists the values of the parameters describing the behaviour of the wall under diagonal compression tests. The mean values, with the respective coefficient of variation, are reported as well. The externally strengthened specimens of the series SW-1L and SW-2L exhibited a mean value of the maximum load, P_{max}, more than two times higher than the value showed by the reference walls of the series USW. However, the two series of reinforced walls, SW-1L and SW-2L, exhibited similar values of the maximum shear capacity, despite having a significantly different amount of textile. This aspect may be explained by analysing the mechanical behaviour of the composite and its components. As it might be expected, based on the Flax TRM tensile tests,

Table 5.2 Diagonal compression test parameters for the specimens of the series USW, SW-1L and SW-2L

Specimen	P_{peak} (kN)	$\tau_{o,peak}$ (MPa)	γ_{peak}(%)	G (MPa)	μ	Γ_1 (J)	Γ_2 (J)
USW-1	66	0.39	0.03	2987	–	9	–
USW-2	73	0.43	0.05	1473	–	11	–
USW-3	53	0.31	0.05	2104	–	11	–
mean	64	0.38	0.04	2188	–	10	–
Co.V.	16	16	27	35	–	11	–
SW-1L-1	164	0.97	0.03	2412	4.25	48	881
SW-1L-2	145	0.85	0.05	1460	3.01	51	1552
SW-1L-3	109	0.64	0.05	3560	4.88	19	1019
mean	139	0.82	0.04	2478	4.05	39	1151
Co.V.	20	20	27	42	23	45	31
SW-2L-1	163	0.96	0.07	2335	3.72	54	1730
SW-2L-2	130	0.77	0.05	1945	3.42	31	2964
SW-2L-3	158	0.93	0.08	2002	4.14	65	3222
mean	150	0.88	0.07	2094	3.76	50	2639
Co.V.	12	12	22	10	10	35	30

in the elastic phase the response of the composite is mainly governed by the mortar matrix, rather than by the reinforcing textile. As seen by the experimental evidence the walls' maximum capacity was attained in the elastic phase, due to the tensile failure of the mortar. The textile threads remained mainly unbroken, except for few well-anchored portions. The textile contribution was fundamental in the post elastic peak phase, when the mortar was already cracked, hence larger displacements allowed the tightening of the fibres.

The shear modulus G, assessed as the shear stress-strain curve slope in the elastic branch, assumed the same order of magnitude for all the specimens of the three series. This parameter is mainly depending on the contribution to the strength of the masonry wall.

The strengthened walls exhibited a significant drop in the load as a consequence of the opening of the cracks along the vertical diagonal. This loss load was of about the 30% of the maximum wall capacity, for the SW-1L specimens, and of about the 15% of the maximum load for the SW-2L specimens. As matter of fact, by doubling the amount of textile, this drop of the load helved. This experimental evidence emphasises that at large values of the strains, in the post peak phase, the fibres provide a significant contribution, and that a higher amount of reinforcing textile can improve the overall post elastic behaviour.

In order to quantify the strain capacity of the system, the coefficient μ was considered, defined as the ratio between the shear strain value corresponding to a loss of bearing capacity equal to 20% of the peak load, in the post peak phase, and the value of the shear strain corresponding to the maximum load.

Unstrengtened walls, not exhibiting a post peak behaviour, were considered not having any ductile capacity. Conversely, SW-1L and SW-2L specimens were characterised by a mean value of the parameter μ varying between 3.8 and 4.0, indicating a capacity of the system of attaining shear strains of about for times larger than the strain at the elastic limit.

The energy dissipated during the test, assessed as the area underlying the load-vertical displacement curve, was adopted as parameter representative of the ductility of the system. Two values of the energy were considered: the energy dissipated in the first branch of the curve until the achievement of the maximum load, labelled as Γ_1, and the energy dissipated during the post peak phase, labelled as Γ_2. Γ_2 was assessed in the range between the peak load and a load equal to the 30% of the peak load. Figure 5.13 shows that by doubling the amount of textile the dissipated energy Γ_2 more than doubled. In line with the other parameters investigated, also in terms of energy the significant contribution of the textile in the post peak phase is confirmed.

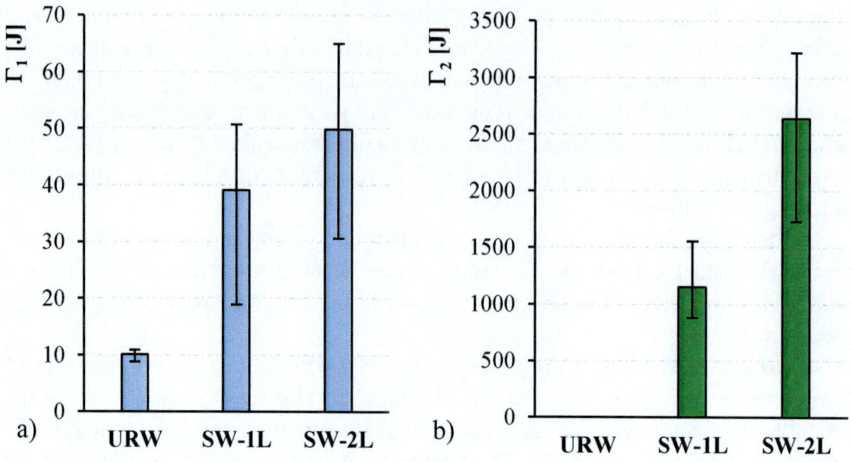

Fig. 5.13 Mean values of the energy dissipated during diagonal compression tests: **a** energy dissipated to achieve the peak load; **b** energy dissipated during the post peak phase

5.3 Comparison with Standards

This section aims at providing a comparison between the shear capacity experimentally observed with the values theoretically assessed via the formulations proposed by standards for the shear capacity of masonry elements without and with the application of the external reinforcement by means of TRM composites. The strength exhibited by the unstrengthened walls, series USW, was compared with the value assessed referring to Chap. 8 of the explanatory circular of the Italian Standard NTC [2], concerning the assessment of the capacity of existing buildings. The increase of shear capacity due to the application of the TRM composite system experimentally observed was compared with the capacity assessed via the formulations provided by the CNR-DT 215/2018 [3], concerning the designing of retrofitting interventions by means of TRMs.

It is worth pointing out that the loading conditions considered within the standards formulation are different from those reproduced by the tests. Notably, the standards provide the shear capacity in direct shear loading conditions, while the experimental analysis was conducted by means of diagonal compression tests. As consequence it is not possible to directly compare the experimental and the theoretical shear capacities. Therefore, for the USW series, the shear stress in the middle of the panel was adopted as parameter for the comparison, for the strengthened walls, SW-1L and SW-2L series, the increase of strength provided by the reinforcement application was adopted as parameter of comparison.

5.3.1 Unstrengthened Walls

The parameter adopted for the comparison of the strength of the USW series is the shear stress τ_0, defined as the maximum shear stress in the middle of the panel in pure shear conditions (in absence of compression stresses). Several approaches can be adopted for the determination of the shear stress τ_0 in diagonal compression loading conditions. According to the formulation proposed by the standard ASTM E519-2 [1], the principal stress directions are oriented along the two diagonals and in the middle of the panel the two principal stresses attain the same value. Under these conditions, Eq. 5.3 define the shear stress in absence of normal stress, $\tau_{0,\mathrm{ASTM}}$, being P_{peak} the maximum load in diagonal compression conditions, and A_n the net area of the wall. The recommendations RILEM TC-76-LUM [4], provide a formulation for the identification of the shear stress in the middle of the panel as the result of an elastic solution implemented with the hypothesis of considering the wall as a homogeneous and isotropic means. With these conditions the two principal stresses in the middle of the panel assumes different values, and the shear stress (corresponding to a null value of the normal stress), $\tau_{0,\mathrm{RILEM}}$, is defined by Eq. 5.4.

$$\tau_{0,ASTM} = \frac{P_{peak}}{\sqrt{2} \cdot A_n} \tag{5.3}$$

$$\tau_{0,RILEM} = \frac{0.90 \cdot P_{peak}}{A_n} \tag{5.4}$$

It is reasonable to assume the experimental value of the maximum shear stress in the centre of the panel to be included between $\tau_{0,\mathrm{ASTM}}$ and $\tau_{0,\mathrm{RILEM}}$. According to the Italian standard code, the theoretical estimation of the masonry element can be assessed from the strength values of the components of the wall (i.e. mortar and bricks). Knowing the mechanical properties of the bricks and of the bed joints mortar and interpolating some reference values proposed by the standard (Table 11.10.IV of the NTC [2]), it is possible to estimate the strength of the wall. According to this procedure the compression strength of the masonry is equal to 6.3 MPa. The standard provides a range of shear stresses in the middle of the panel, $\tau_{0,1} \div \tau_{0,2}$, as a function of the wall compression strength and of the masonry typology (Table C8.5.I and Table C8.5.II of the NTC [2], Chap. 8 of the explanatory circular). Table 5.3 lists the shear stress range values deriving from both experimental and theoretical approaches. The shear stress τ_0 experimentally assessed assumes a value of about 2.5 times higher than the average theoretical value. This aspect emphasises the wide safety margin of the standard code in proposing a value of strength for design purposes, in absence

Table 5.3 Shear strength of unstrengthened walls: experimental and theoretical results

Experimental results			Standard model
$\tau_{0,\mathrm{ASTM}}$	0.38	MPa	0.10 MPa < τ_0 < 0.25 MPa
$\tau_{0,\mathrm{RILEM}}$	0.48	MPa	

of experimental data. It is also highlighted that the pure shear conditions are not perfectly guaranteed by the diagonal compression test loading configuration.

The modified formulation of Turnsek-Cacovic for the assessment of the shear capacity of the unreinforced wall, in direct shear condition, is proposed by the Italian standard NTC [2]. In absence of normal stresses this formulation assumes the form of the Eq. 5.5:

$$V_t = \frac{A_n \cdot 1.5 \cdot \tau_0}{b} \tag{5.5}$$

in which A_n represents the net section of the wall (product of the length and the thickness), τ_0, defined as the mean value of the range $\tau_{0,1} \div \tau_{0,2}$, is the shear stress in absence of normal stresses, b is a corrective coefficient for taking into account the distribution of the stress through the section (assumed equal to 1). The shear capacity of the unreinforced wall, V_t, according to the theoretical formulation, and in direct shear conditions is equal to 31 kN.

5.3.2 TRM Externally Strengthened Walls

The theoretical shear capacity of reinforced walls was computed to compare the increase of strength with the one observed experimentally. The shear capacity was assessed according to the Italian instructions for designing TRM interventions Linea Guida [5]. The masonry panel is considered externally strengthened by TRM composites, uniformly all over the surface and symmetrically on both the sides, and with the fibres oriented n both vertical and horizontal directions. The shear capacity of masonry walls loaded in their plane, $V_{t,R}$, is the sum of the contribution of the unstrengthened wall, V_t, and of the TRM composite, V_{tf}. The contribution of the composite can be assessed according to Eq. 5.6:

$$V_{t,f} = 0.5 \cdot n_f \cdot t_{v,f} \cdot b_f \cdot \alpha_t \cdot \varepsilon_{f,d} \cdot E_f \tag{5.6}$$

in which product $n_f \cdot t_{v,f} \cdot b_f$ is the area of the equivalent section of the actual shear reinforcement oriented parallel to the shear force; α_t is a coefficient that takes into account the decrease of tensile strength of the textile when subjected to shear (it is assumed equal to 0.8); ε_{fd} represents the maximum deformation of the textile superiorly limited by the maximum value attained during the shear bond test; E_f represents the stiffness of the dry textile. The strain was assumed to be equal to ε^*, whose assessment procedure is shown in paragraph Eq. 5.6 (Chap. 4).

The contribution to the shear capacity of the composite systems Flax-TRM-1L and Flax-TRM-2L, $V_{t,f}$, assessed with applying Eq. 5.6 was respectively equal to 19 kN and 38 kN. The formulation proposed by the standard consider this contribution to be linearly dependent from the amount of textile in the composite system. The

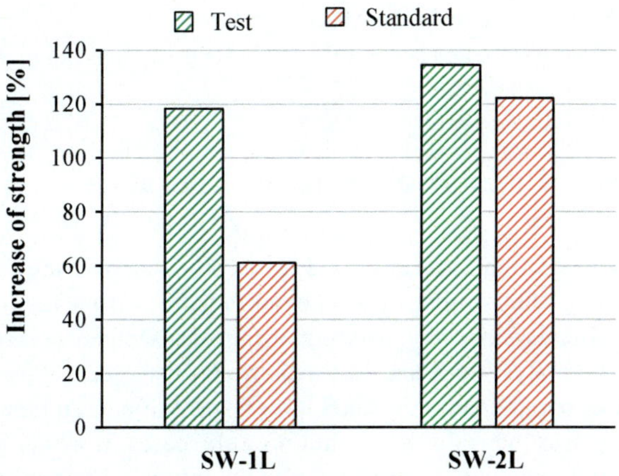

Fig. 5.14 Comparison in terms of increases of strength deriving from the standard models and the experimental tests

global shear capacity of the walls, $V_{t,R}$, obtained by summing the contribution of the composites, $V_{t,f}$, and the contribution of the unstrengthened wall, V_t, assumed values equal to 50 kN and 69 kN respectively for the SW-1L and the SW-2L configurations. Figure 5.14 shows the ratio between the capacity of the strengthened wall and the reference capacity of the unstrengthened panels.

With respect to the series SW-1L, a significant discrepancy was observed between the increase of strength experimentally assessed and the one theoretically computed. The difference between the experimental and analytical results significantly decreases in the case of the SW-2L series. In both the cases the model proposed by the standard provides lover increase of strength values. Moreover, contrary to what proposed by the standard model, the increase of strength by changing the amount of textile is not proportional to the fibre textile reinforcement. This result emphasises that an increase of reinforcement obtained by applying several textile plies, does not necessary results in the increase of strength obtainable by using a denser textile grid. As matter of fact, the global behaviour of the system seems to be affected by the textile arrangement, especially in the case of the use of plant fibres, characterised by a non-uniform distribution of the stress through the textile due to the unavoidable imperfections of the natural fabric itself.

The formulation proposed by the standard is based on the assumption of considering the maximum capacity of the composite coinciding with the tensile strength of the reinforcing textile. In the present study, although the rupture of the textile was attained at the ultimate state, the maximum load was achieved at the end of the elastic phase, with the failure in tension of the mortar matrix, and with the fibres mainly unbroken. Therefore, the discrepancy between the experimental and theoretical results may be attributed to the limit of the standard model not considering failure modes not involving the rupture of the textile. As matter of fact, in the formulas proposed by the standards, the nature, hence the mechanical properties, of the mortar matrix was never considered. Probably, further analysis aimed at obtaining a more sophisticated formulation, that would better consider the contribution of all the components of the TRM-to-masonry system, may provide a more efficient model capable to better predict the global mechanical behaviour.

5.4 Comparison with the Literature

This section aims at providing a comparison of the results obtained in the present study with similar experimental outcomes from the literature concerning diagonal compression tests carried out on masonry elements externally strengthened by TRMs. The analysis was carried out on two levels: a general comparison of the main mechanical parameters performed with respect to a large heterogeneous database of walls, by also focusing the attention on the cases in which plants fibres were adopted as reinforcement; a comparison with similar studies in which reinforcement configurations obtained by overlapping several textile plies were considered. In order to have a large database, unavoidably walls characterised by different building blocks, geometry and reinforcement system with respect to what presented in this study, were considered as comparison. Only studies concerning diagonal compression tests carried out in laboratory were considered, excluding in situ tests, in order have a uniformity in the test set-up. To take into account this heterogeneity, for each series of specimens considered, the main characteristics, such as wall size, masonry type, TRM adopted and reinforcement configuration, were listed in Table 5.4. As reported in Table 5.5, several parameters to describe the mechanical properties of the specimens were considered as well, such as the strength of both the mortar and the textile adopted in the TRM system and the reinforcement rate (defined as the maximum capacity in tension of the textile included in a TRM strip one meter wide, expressed in kN/m). As proposed in similar studies [6] the mortar mechanical ratio, ω_m and the grid mechanical ratio, ω_f, representing an index of the contribution in terms of strength, respectively of the TRM mortar and textile, were considered as well and computed according to Eqs. 5.7 and 5.8:

$$\omega_m = \frac{f_t \cdot A_f}{\tau_0 \cdot A_{masonry}} \tag{5.7}$$

$$\omega_f = \frac{\sqrt{f_{cm}} \cdot A_{matrix}}{\tau_0 \cdot A_{masonry}} \tag{5.8}$$

where f_t is the ultimate tensile strength of the textile grid, τ_0 represents the capacity in terms of shear stress of the unstrengthened configuration, $A_{masonry}$ represents the wall section (L·t, as defined in Fig. 5.3 f_{cm} is the compression strength of the TRM mortar, and A_{matrix} represents the mortar matrix section ($L \cdot t_{matrix}$), A_f is the total area of the mesh reinforcement computed as defined by [6] (it is the product among: $n_{orientation}$, a coefficient related to the grid orientation equal to 1 for unidirectional fabric with angle 0°, equal to 2 for bidirectional grids with orientation angle 0°–90°;

Table 5.4 Comparison of test results with data from the literature for TRM strengthened masonry panels in Diagonal Compression

Paper	Wall size mm^3	Mas. type	TRM textile	TRM mortar	$f_{c,m}$ MPa	t_{TRM} mm	Reinf. conf.	$f_{t,tex}$ MPa	ρ kN/m	τ_{max} MPa	$\frac{\tau_{max}}{\tau_{USW}}$	G MPa	$\frac{G}{G_{USW}}$	μ	Failure mode	ω_f	ω_m
SW-1L series	1000 × 1000 × 120	CB	Flax	HL	11.1	5	Unif.	354	31	0.82	2.2	2478	1.1	4.1	C-TRM, CS-Deb	2.7	0.7
SW-2L series	1000 × 1000 × 120	CB	Flax	HL	11.1	8	Unif. (2 pi.)	354	62	0.88	2.3	2095	1.0	4.4	C-TRM, CS-TRM	5.4	1.2
Menna et al. [7]	1200 × 1200 × 250	TB	Hemp	PM	16.1	15	Unif*	–	20	0.50	1.9	601	1.26	5.1	C-TRM, CS-TRM	1.2	1.8
	1180 × 1180 × 250	CB	Hemp	PM	16.1	15	Unif*	–	20	0.75	5.3	445	1.21	7.4	C-TRM	2.3	3.4
	1200 × 1200 × 250	TB	Hemp	HL	14.2	40	Unif*	–	20	0.75	2.8	1543	3.24	7.1	C-TRM	1.2	4.5
	1180 × 1180 × 250	CB	Hemp	HL	14.2	40	Unif*	–	20	0.76	5.4	2250	6.14	32.2	C-TRM	2.3	8.6
Olivito et al. [8]	510 × 510 × 120	CB	Hemp	–	–	–	Grid	240	27	1.34	1.2	3065	1.32	–	CS-TRM	0.8	–
	510 × 510 × 120	CB	Hemp	–	–	–	Grid	240	53	1.32	1.2	3096	1.33	–	CS-TRM	1.6	–
	510 × 510 × 120	CB	Flax	–	–	–	Grid	350	54	2.13	2.0	2870	1.23	–	CS-TRM	1.7	–
	510 × 510 × 120	CB	Flax	–	–	–	Diag	350	26	1.77	1.6	3043	1.31	–	C-TRM	0.8	–
	510 × 510 × 120	CB	Flax	–	–	–	Diag	350	13	1.56	1.5	3042	1.31	–	C-TRM	0.4	–
	510 × 510 × 120	CB	Glass	–	–	–	Grid	–	–	1.53	1.4	4128	1.77	–	CS-TRM	–	–

(continued)

Table 5.4 (continued)

Paper	Wall size mm^3	Mas. type	TRM textile	TRM mortar	$f_{c,m}$ MPa	t_{TRM} mm	Reinf. conf.	$f_{t,tex}$ MPa	ρ kN/m	τ_{max} MPa	$\frac{\tau_{max}}{\tau_{USW}}$	G MPa	$\frac{G}{G_{USW}}$	μ	Failure mode	ω_f	ω_m
	510 × 510 × 120	CB	Glass	–	–	–	Diag	–	–	1.54	1.4	3772	1.62	–	C-TRM	–	–
Babaeidarabad et al. [9]	1200 × 1200 × 120	CB	Carbon	Cem	20.0	10	Unif	802	40	1.32	2.2	1520	1.8	18.2	C-TRM	2.2	1.2
	1200 × 1200 × 120	CB	Carbon	Cem	20.0	10	Unif (4 pi.)	802	160	2.86	4.7	2660	3.2	4.3	TC	8.8	1.2
Marcari et al. [10]	1000 × 1000 × 250	TB	Basalt	HL	15.0	6	Unif	3000	110	0.62	1.6	1017	1.4	6.7	C-TRM	4.6	0.5
Wang et al. [11]	910 × 910 × 110	CB	Carbon	Cem	32.0	15	Grid	3110	56	1.80	2.1	2806	1.6	4.3	C-TRM, TC	2.4	1.8
	910 × 910 × 110	CB	Steel	Cem	32.0	10	Grid	2800	72	1.50	1.8	2300	1.3	14.9	TC	3.1	1.2
	910 × 910 × 110	CB	Basalt	Cem	32.0	15	Unif (2 pi.)	1006	35	1.67	2.0	2529	1.5	1.0	C-TRM, CS-TRM	1.5	1.8
	910 × 910 × 110	CB	Basalt	Cem	32.0	15	Unif (4 pi.)	1006	70	1.71	2.0	4538	2.6	13.4	TC	3.0	1.8
Almeida et al. [12]	990 × 990 × 140	HCB	Carbon	Cem	44.0	25	Unif	4300	185	2.09	4.2	4458	1.6	–	C-TRM, CS-Deb	11	4.8
Parisi et al. [13]	1230 × 1230 × 310	TB	GFRP	HL	16.1	10	Unif	1270	51	0.65	3.0	570	1.4	5.4	C-TRM	3.1	1.2

(continued)

Table 5.4 (continued)

Paper	Wall size mm^3	Mas. type	TRM textile	TRM mortar	$f_{c,m}$ MPa	t_{TRM} mm	Reinf. conf.	$f_{t,tex}$ MPa	ρ kN/m	τ_{max} MPa	$\frac{\tau_{max}}{\tau_{USW}}$	G MPa	$\frac{G}{G_{USW}}$	μ	Failure mode	ω_f	ω_m
Gattesco et al. [14]	1160 × 1160 × 250	CB	GFRP	HL-Cem	6.7	30	Unif	490	168	0.93	2.0	3481	1.64	16.8	C-TRM	5.8	1.3
	1160 × 1160 × 250	CB	GFRP	HL-Cem	6.7	30	Unif	490	84	0.97	2.1	3146	1.48	7.8	C-TRM	2.9	1.3
	1160 × 1160 × 250	CB	GFRP	HL-Cem	6.7	30	Unif	490	56	1.00	2.1	2953	1.40	4.7	C-TRM	1.9	1.3
	1160 × 1160 × 250	CB	GFRP	HL-Cem	6.7	30	Unif	490	168	1.02	1.6	5165	1.51	15.5	C-TRM	4.3	1.0
	1160 × 1160 × 250	CB	GFRP	HL-Cem	6.7	30	Unif	490	84	1.20	1.9	4783	1.40	5.3	C-TRM	2.1	1.0
	1160 × 1161 × 250	CB	GFRP	HL-Cem	6.7	30	Unif	490	56	1.23	2.0	4513	1.32	4.9	C-TRM	1.4	1.0
	1160 × 1161 × 380	CB	GFRP	HL-Cem	6.7	30	Unif	490	168	0.77	1.7	3843	1.48	15.7	C-TRM	3.9	0.9
	1160 × 1161 × 380	CB	GFRP	HL-Cem	6.7	30	Unif	490	84	0.77	1.7	3506	1.35	5.3	C-TRM	1.9	0.9

(continued)

Table 5.4 (continued)

Paper	Wall size mm^3	Mas. type	TRM textile	TRM mortar	$f_{c,m}$ MPa	t_{TRM} mm	Reinf. conf.	$f_{t,tex}$ MPa	ρ kN/m	τ_{max} MPa	$\frac{\tau_{max}}{\tau_{USW}}$	G MPa	$\frac{G}{G_{USW}}$	μ	Failure mode	ω_f	ω_m
	1160 × 1161 × 380	CB	GFRP	HL-Cem	6.7	30	Unif	495	105	0.90	2.0	4258	1.64	11.0	C-TRM	2.4	0.9
	1160 × 1161 × 380	CB	GFRP	HL-Cem	6.7	30	Unif	495	70	0.70	1.5	3984	1.54	8.4	C-TRM	1.6	0.9
Ferretti et al. [15]	1200 × 1200 × 120	CB	Carbon	FR-PM	6.5	6	Unif*	2363	111	1.17	1.9	–	–	–	C-TRM	6.1	0.4
	1200 × 1200 × 120	CB	Glass	FR-PM	6.5	6	Unif*	991	63	1.24	2.1	–	–	–	C-TRM	3.5	0.4
Incerti et al. [16]	1100 × 1100 × 250	CB	Basalt	HL	12.6	6	Unif*	1016	33	0.85	1.5	2575	1.02	–	C-TRM	0.9	0.3
	1160 × 1160 × 250	CB	Basalt	HL	12.6	6	Unif*	1016	33	0.79	1.9	2537	1.71	–	C-TRM	1.3	0.4
Del Zoppo et al. [17]	1280 × 1280 × 140	CB	BFRP	FR-HL	14.5	15	Unif	1542	60	1.28	3.7	2350	1.38	5.0	C-TRM	4.9	2.3
	1280 × 1280 × 285	CB	BFRP	FR-HL	14.5	15	Unif	1542	60	0.90	1.6	2857	1.01	6.8	C-TRM	1.5	0.7

(continued)

Table 5.4 (continued)

Paper	Wall size mm^3	Mas. type	TRM textile	TRM mortar	$f_{c,m}$ MPa	t_{TRM} mm	Reinf. conf.	$f_{t,tex}$ MPa	ρ kN/m	τ_{max} MPa	$\frac{\tau_{max}}{\tau_{USW}}$	G MPa	$\frac{G}{G_{USW}}$	μ	Failure mode	ω_f	ω_m
De Lorenzis et al. [18]	461 × 461 × 100	CS	Carbon	PM	20.4	7	Unif	2800	132	1.75	1.2	–	–	–	C-TRM	3.7	0.4
Prota et al. [19]	1030 × 1030 × 250	TB	Glass	FR-Cem	24.1	10	Unif (2 pi.)	1276	87	0.59	2.5	731	1.43	3.3	C-TRM	5.8	1.6
	1030 × 1030 × 250	TB	Glass	FR-Cem	24.1	10	Unif	1276	44	0.50	2.1	492	1.00	3.5	CS-TRM	2.9	1.6
Faella et al. [20]	1160 × 1160 × 380	TB	Carbon	FR-Cem	38.0	10	Unif	4800	160	0.33	5.6	–	–	–	C-TRM, C-Deb	29	5.5
Balsamo et al. [21]	1000 × 1000 × 250	TB	GFRP	FR-PM	16.1	10	Unif	–	45	0.52	3.3	821	2.96	4.7	C-TRM	4.6	2.1
	1000 × 1000 × 250	TB	BFRP	FR-PM	16.1	10	Unif	–	60	0.39	2.5	369	1.33	2.4	C-TRM	6.2	2.1
	1000 × 1000 × 250	TB	GFRP	Cem	28.3	10	Unif	–	45	0.64	4.1	928	3.35	4.9	C-TRM	4.6	2.7
Del Zoppo et al. [22]	1200 × 1200 × 250	TB	Steel	Cem	18.8	40	Unif*	451	115	0.97	3.8	2338	3.93	1.1	C-TRM, C-Deb	7.2	5.4
	1200 × 1200 × 250	TB	Steel	FR-Cem	14.2	40	Unif*	451	115	0.93	3.6	2196	3.69	1.9	C-TRM, C-Deb	7.1	4.7
	1200 × 1200 × 250	TB	Steel	FR-HL	14.5	40	Unif*	451	115	0.76	2.9	1552	2.61	1.5	C-TRM, C-Deb	7.1	4.7
	1200 × 1200 × 250	TB	GFRP	FR-Cem	14.2	40	Unif*	1350	76	0.94	3.7	1857	3.12	2.7	C-TRM, C-Deb	4.7	4.7
	1200 × 1200 × 250	TB	GFRP	FR-HL	15	40	Unif*	1350	76	0.62	2.4	1646	2.77	2.2	C-TRM, C-Deb	4.7	4.8

(continued)

Table 5.4 (continued)

Paper	Wall size mm^3	Mas. type	TRM textile	TRM mortar	$f_{c,m}$ MPa	t_{TRM} mm	Reinf. conf.	$f_{t,tex}$ MPa	ρ kN/m	τ_{max} MPa	$\frac{\tau_{max}}{\tau_{USW}}$	G MPa	$\frac{G}{G_{USW}}$	μ	Failure mode	ω_f	ω_m
	1200 × 1200 × 250	TB	GFRP	FR-Cem	14	40	Unif*	1480	53	1.02	4.0	2252	3.79	2.0	C-TRM, C-Deb	3.3	4.7
	1200 × 1200 × 250	TB	GFRP	FR-HL	15	40	Unif*	1480	53	0.78	3.0	1384	2.33	4.6	C-TRM, C-Deb	3.3	4.7
Babaeidarabad et al. [23]	1220 × 1220 × 92	ConB	Carbon	Cem	20.0	10	Unif	802	38	2.06	1.9	3747	1.22	3.2	TC	1.6	0.9
	1220 × 1220 × 92	ConB	Carbon	Cem	20.0	10	Unif (4 pi.)	802	154	2.49	2.4	4940	1.61	1.3	TC	6.3	0.9
Ismail et al. [24]	1050 × 1050 × 150	ConB	GFRP	FR-PM	40	10	Unif	–	105	1.23	2.5	18415	5.23	12.0	C-TRM	5.6	1.7
	1050 × 1050 × 150	ConB	Basalt	FR-PM	39	10	Unif	–	60	1.35	2.7	13190	3.74	13.2	C-TRM	3.2	1.7
	1050 × 1050 × 150	ConB	CFRP	FR-PM	51	10	Unif	–	240	1.10	2.2	15050	4.27	15.1	TC	13	1.9

Table 5.5 Nomenclature adopted in Table 5.4

Properties	
$f_{c,m}$	Compressive strength of TRM mortar
t_{TRM}	TRM thickness
Reinf. conf.	TRM reinforcement scheme
$f_{t,tex.}$	Tensile strength of TRM textile
ρ	Tensile strength of the textile per unit wigth
τ_{max}	Maximum shear stress in the centre of the panel evaluated according to ASTME519
τ_{max}/τ_{USW}	Ratio between the maximum shear stress in the reinforced configuration over the maximum shear stress of the reference
G	Elastic shear modulus evaluated according to the ASTM E519
G/G_{USW}	Ratio between the shear modulus in the reinforced configuration over the shear modulus of the reference unstrengthened wall
μ	Ductility
ω_f	Grid mechanical ratio
ω_m	Matrix mechanical ratio
Masonry types	
CB	Clay brick units
TB	Tuff blocks
HCB	Holed ceramic bricks
CS	Calcareous stones
ConB	Concrete blocks
TRM mortar	
HL	Hydraulic lime mortar
PM	Pozzolanic mortar
Cem	Cement-based mortar
FR	Short fiber reinfroced mortar
Reinf. Conf.	
Unif (x pl.)	TRM uniformly applied over all the surface number of overlapped textile plies in the TRM (in the absence of number, a single ply is implied)
Grid	Reinforcement with TRM strips oriented parallel to the wall sides (45° with respect to the load direction)
Diag	Reinforcement with TRM strips oriented along the two wall diagonals (0°–90° with respect to the load direction)
*	Use of mechanical anchors
Failure mode	
C	Loaded diagonal direction
CS	Along the mortar joints
TRM	TRMfailure, either for textile slippage or textile rupture
Deb	Debonding at the masonry-TRMinterface
TC	Toe crushing at the loaded edges

n_{side}, the number of strengthened sides of the wall; n_{layers}, the number of grid layers in each side; t_{eff}, the grid equivalent dry thickness; L, the wall edge length).

In order to compare the mechanical response under diagonal compression tests the following parameters were considered: the maximum capacity of the wall in terms of shear stress, τ_{max}, and the shear modulus, G, both defined according to the ASTM E519-2 [1], the ratio between their values in the strengthened configuration over the values concerning the unstrengthened reference one, $\tau_{max}/\ \tau_{USW}$ and G/G_{USW}, the ductility, μ, computed as the ratio γ_u/γ_{el} between the shear strain at the elastic limit and at the ultimate configuration (γ_{el} corresponds to the furthermost point at which shear stress and strain are linearly correlated, depending on the shear stress-strain curve it corresponds either to the end of the elastic branch or to the point of the elastic phase corresponding to the 80% of the peak load, γ_u typically corresponds to the 80% of the peak load in the post peak phase).

5.4.1 Description of the Database

Several studies exist in literature concerning the mechanical characterisation of the shear capacity of masonry elements externally strengthened by TRM composite systems. Among them only those complying with some criteria imposed by the author were considered in order to be consistent with the experimental activity herein presented. Nevertheless, the database still resulted to be quite heterogeneous due to the intention of having a large amount of data.

All studies considered concerned externally strengthened walls subjected to diagonal compression test according to the ASTM E519-2 [1]. The walls taken into consideration included mostly clay brick masonry, both in single and double leaf configurations, and in some cases tuff blocks or other less common blocks masonry elements.

Concerning the TRM system the most common industrial fibres were considered, such as glass, carbon, basalt and steel; GFRP and CFRP reinforcement grids were considered too, as well as few cases in which plant fibres, such as hemp and flax, were employed.

The matrices adopted were mainly either cement-, hydraulic lime- or pozzolanic-based. In all cases, the TRM reinforcement was symmetrically applied on both sides of the wall, and almost in all cases the reinforcement was uniformly applied all over the surface. In few cases different TRM reinforcement configurations, performed by the use of TRM strips or by the use of mechanical anchors, were considered as well (in these studies plant fibres were adopted as reinforcement in TRMs).

Concerning the failure mode five different cases were considered: toe crushing at the loaded edges, cracking of masonry along the loaded diagonal, combined cracking of masonry and sliding along the mortar joints, TRM failure either due to textile slippage or rupture, debonding at the TRM to masonry interface. In some cases, the failure mode resulted a combination of the just mentioned modes.

5.4.2 Comparison with the Entire Database

The comparison was carried out by analysing the relationship between the mechanical parameters above mentioned with the grid and mortar mechanical ratios, ω_m and ω_f. Each one of the parameters τ_{max}/τ_{USW}, G/G_{USW} and μ was studied individually by distinguishing the case studied herein from the cases in which high strength fibre textile and plant fibre fabrics were adopted.

In Fig. 5.15a and b the relationship between τ_{max}/τ_{USW} and respectively, ω_m and ω_f is shown. As already observed in similar studies [6], the graphs confirmed the enhancement of strength capacity to be mainly depending on the mortar contribution rather than on the fibres one. Moreover, the graphs showed that plant fibre TRM based systems were generally characterised by lower values of the grid mechanical ratio, ω_f. This aspect is mainly due to the lower strength of plant fibres with respect to the industrial ones. It is worth emphasising that the only value out of this range is the one concerning the author's case of study characterised by the double plies flax textile reinforcing configuration.

The same comparison was proposed in terms of ductility, μ, in Fig. 5.16a and b. Although in both cases the data do not follow a clear trend, an increase of the ductility was observed in the range of greater values of the grid mechanical ratio, ω_f. Due to the low values of the grid mechanical ratio, ω_f, no significant information could be observed in terms relationship between ductility and amount of plant fibres. In the author's case of study, the grid mechanical ratio did not significantly affect the parameter of ductility, μ.

The comparison in terms of G/G_{USW} (Fig. 5.17a and b) showed that by increasing the mortar contribution, ω_m, a quasi linear increase of the shear modulus was

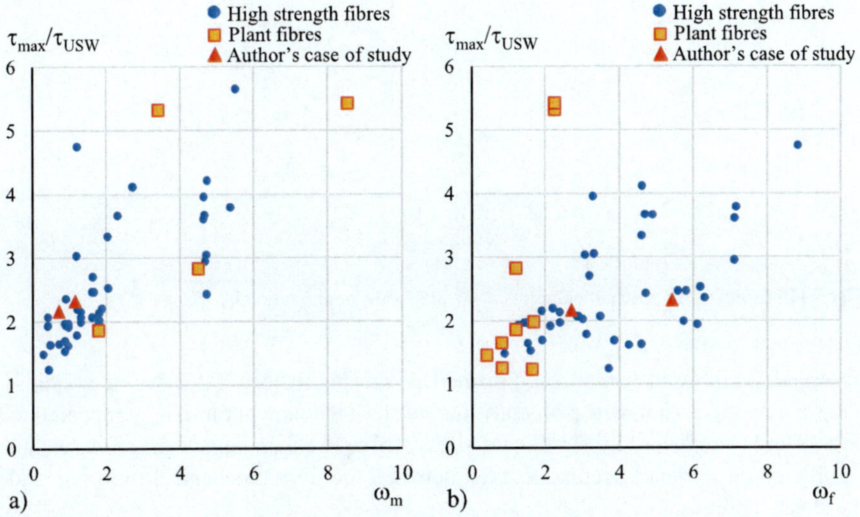

Fig. 5.15 Effect of the mechanical ratios **a** ω_m and **b** ω_f, on the ratio τ_{max}/τ_{USW}

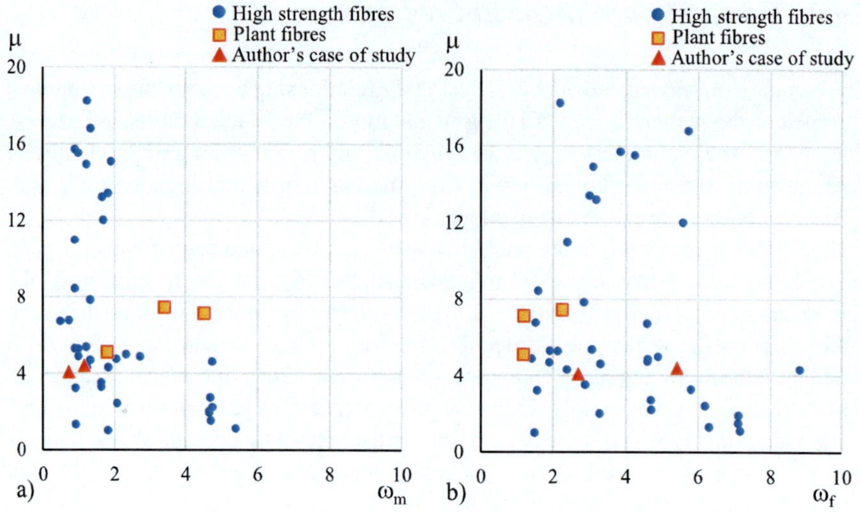

Fig. 5.16 Effect of the mechanical ratios **a** ω_m and **b** ω_f, on the ductility μ

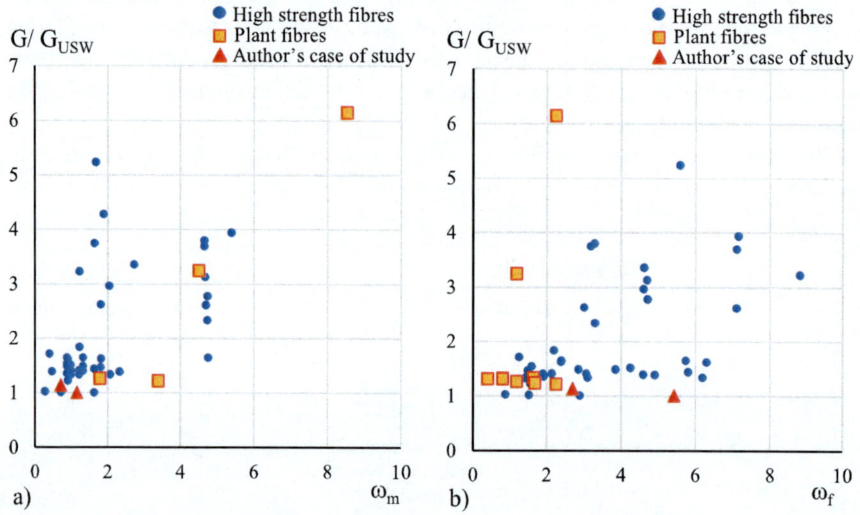

Fig. 5.17 Effect of the mechanical ratios **a** ω_m and **b** ω_f, on the ratio G_{max}/G_{USW}

observed. This trend was also confirmed in the plant fibre TRM based systems. Concerning the author's case of study, the mortar mechanical ratio, ω_m, appeared to be too low to register any significant shear stiffness enhancement. In all cases, no significant dependence seemed to exist between the fibre mechanical ratio, ω_f, and the shear modulus, G.

This aspect, together with the outcomes deriving from the comparison in terms of shear stress, τ_{max}, highlighted that the elastic behaviour of masonry walls externally strengthened by TRM composite systems, is mainly governed by the strength contribution provided by the TRM mortar matrix, rather than by that one deriving from the fibre reinforcing grid. Such evidence was expected, it is known that fibres are activated once the mortar (and the wall) is cracked.

5.4.3 Comparison Between Walls Strengthened by Multi-ply Grids

Aiming at enhancing the shear capacity of TRM externally bonded systems by increasing the fibre mechanical ratio, ω_f, several studies, as well as the author's one, proposed a TRM composite system by applying overlapped textile plies (herein labelled as multi-ply grids technique), taking care that a thin layer of mortar is always introduced between the different textile grids. In order to assess the influence of the multi-ply grid technique on the mechanical behaviour of TRM externally strengthened walls subjected to diagonal compression test, the author's case of study was compared with other four cases of study in which the same reinforcement technique was adopted. Specifically, in two studies, similar to what is proposed in the thesis, the textile amount was doubled [11, 19], in other two studies it was quadrupled [23, 9]. The comparison was carried out in terms of maximum shear stress, τ_{max}, shear modulus, G and ductility, μ. The ratio between these parameters concerning the multi-ply configuration (mp) and the same parameters concerning the so-called standard-ply (sp) arrangement (τ_{mp}/τ_{sp}, G_{mp}/G_{sp}, μ_{mp}/μ_{sp}) are listed, for each case of study, in Table 5.6.

In terms of maximum shear stress, τ_{max}, in the current study it was observed an increase of the 7% by doubling the amount of textile. This experimental evidence was in line with what observed in similar studies in which the 2-plies configuration was adopted. With respect to the case of the study Babaeidarabad et al. [9], a significant enhancement of strength was observed by increasing the percentage of textile of the 300%. The same gain of strength was not attained in the study of [23] in which the 4-plies configuration was adopted as wall. This discrepancy may be due to the fact

Table 5.6 Comparison between walls externally strengthened by TRMs performed with the multi-ply grid technique

Case of study	Increase of plies	τ_{mp}/τ_{sp}	G_{mp}/G_{sp}	μ_{mp}/μ_{sp}
Author's case of study	×2	1.07	0.85	1.1
Wang et al. [11]	×2	1.02	1.79	13.4
Prota et al. [19]	×2	1.19	1.49	0.9
Babaeidarabad et al. [9]	×4	2.17	1.75	0.2
Babaeidarabad et al. [23]	×4	1.21	1.32	0.4

that in the study of [23], a stiffer masonry was adopted (see Table 5.4): the smaller is the strain in the elastic phase, the smaller is the fibres contribution in this phase. In terms of elastic shear modulus, G, in all the cases of comparison an increase of stiffness was recorded, with increments in the range of the 30% to the 80%. On the contrary, in the author's case of study, the higher amount of textile lead to even lower values of the shear modulus. Regarding the ductility, μ, except for a value largely outside the mean trend [11], no important benefits were observed.

As a matter of fact, with the limitation of an analysis carried out on a small set of data, it could be affirmed that by overlapping several textile plies, no significant benefits were conferred to the system. This result may be due to the fact that the multi-ply reinforcing technique did not always guarantee the textile to be uniformly charged through the entire surface subjected to a stress concentration [25].

Moreover, it is worth underlining that the entire analysis was carried out without taking into account the reinforcing textile stiffness in tension. In the author's specific case of study, for example, the structure of the flax textile did not allow the grid to be tied, in a proper uniform way, during the implementation of the TRM composite system. This may be the reason why, during the elastic phase, the strain level resulted to be not enough to let the textile immediately contribute to the whole system strength. The poor capacity to uniformly distribute the stress through the entire textile is a typical aspect of non-made fibres, even more evident in the case of plant fibre textiles due to the morphology of the natural threads conferring to the fibres lower stiffness at low strain values [26, 27].

With a view to future applications it may be interesting to investigate the increase of shear capacity of externally strengthened masonry panels by means of Flax-TRM composite systems, by varying the mortar mechanical ratio, ω_m, i.e. by adopting mortars with higher strength or by increasing the TRM thickness. It may be interesting also to investigate systems in which higher values of the fibre mechanical ratio, ω_f, are obtained by using specifically designed grids. This has to be done by increasing the specific surface of the textile in contact with the matrix, ensuring at the same time a good degree of mortar penetration trough the openings of the textile. Finally, impregnation treatments may make the textile tighter, facilitating its application within the fresh mortar, and so by adopting a better collaboration between textile and mortar also on smaller level of deformation.

5.5 Conclusions

This chapter aimed at investigating the efficiency of the studied Flax TRM systems as strengthening of structural masonry elements. The experimental results were compared with the values of strength computed according to the model proposed by standards in the field. The main outcomes are shown as follow:

- the mechanical response of clay-brick masonry walls externally strengthened by Flax-TRM systems was significantly affected by the presence of the reinforcement. An increase of the maximum shear capacity of the 118% was observed in the specimens characterised by the one layer reinforcing configuration, SW-1L, and an increase of the 136% was observed in the case of the system characterised by two overlapped flax textile plies, SW-2L;
- the walls failed due to the development of cracks along the loading vertical diagonal, and due to a gradual tensile rupture of the fibres with the widening of the cracks. No significant debonding phenomena occurred at the TRM-to-masonry interface surface. Moreover, the Flax TRM conferred a "pseud-ductility" to the entire system, consisting in a post-failure integrity of the wall, as shown by the presence of a post-peak phase that was completely absent in the brittle behaviour of the unstrengthened walls;
- the two strengthened configurations, SW-1L and SW-2L, showed a similar maximum shear stress $\tau_{o,max}$, however the energy dissipated in the post-peak phase, Γ_2, by the SW-2L specimens was more than two times higher than the energy dissipated by the SW-1L specimens;
- with respect to the unstrengthened walls, USW, the shear stress in absence of normal stress, τ_0, deriving from the experimental tests, attained much higher values with respect to the one assessed by the theoretical formulation proposed by standards in the fields, confirming the significant safety margin of the latter;
- regarding the externally strengthened specimens, SW-1L and SW-2L, the increase of strength theoretically assessed is conservative with respect to the experimental evidence. In addition, unlike the model proposed by the standard, the experimental investigation showed that the increase of strength is not directly proportional to the increase of the amount of textile within the composite;
- the comparison with the literature shows that the results are quite in line with similar studies, although showing a low value of the mortar mechanical ratio with respect to systems in which high strength fibres are adopted. The experimental evidences confirm that the maximum shear stress, τ_{0max}, and the shear modulus, G, mainly depend on the mortar mechanical ratio, ω_m, rather than on the fibre mechanical ratio, ω_f; whereas regarding the ductility, μ, no specific trend is observed with either of the two mechanical parameters;
- the comparison with studies in which the multi-ply technique was adopted shows that, in line with the other results, no significant increase of both maximum shear stress, τ_{0max}, and ductility, μ, is observed by overlapping several textile plies. As regard the shear modulus, G, unlike the evidence from other studies, in the present study no increase of the stiffness is observed by doubling the amount of textile.

As a conclusion, the experimental investigation confirmed the potential in the use of plant fibres, i.e. flax textile, as reinforcement in TRMs to enhance the shear capacity of masonry assemblages. However, it also emphasized some limitations in the use of such technique, mainly related to the bond properties between fibres and mortar, the amount of either textile and/or mortar employed, the capacity of the textile to uniformly distribute the stress through all the stressed section. Therefore, there is still room for improvements, by studying possible solutions to solve the mentioned issues, perhaps by investigating fibre impregnation processes to let the grid respond in a more uniform way, or by conducting parametric studies by varying the textile amount to reach higher capacities in the post peak phase.

References

1. ASTM E519-2: 2003 (2003). Standard test method for diagonal tension (shear) in masonry assemblages. In: ASTM Committee C15 on manufactured masonry units, West Conshohocken, PA, United States
2. NTC 2018 (2018) Norme Tecniche per le Costruzioni. In: Ministry of the infrastructures and transports of the Italian Republic
3. CNR-DT 215/2018 (2018) Istruzioni per la Progettazione, l'Esecuzione ed il Controllo di Interventi di Consolidamento Statico mediante l'utilizzo di Compositi Fibrorinforzati a matrice inorganica. In: Consiglio Nazionale delle Ricerche–Commissione per lo studio per la predisposizione e l'analisi di norme tecniche relative alle costruzioni (Italy)
4. RILEM TC-76-LUM (1994) Diagonal tensile strength tests of small wall specimens, RILEM recommendations for the testing and use of constructions materials, 488–489
5. Linea Guida per la identificazione, la qualificazione ed il controllo di accettazione di compositi fibrorinforzati a matrice inorganica (FRCM) da utilizzarsi per il consolidamento strutturale di costruzioni esistenti. In: Consiglio Superiore Lavori Pubblici (Italy) (2018)
6. Del Zoppo M, Di Ludovico M, Prota A (2019) Analysis of FRCM and CRM parameters for the in-plane shear strengthening of different URM types. Composit Part B Eng 171:20–33
7. Menna C, Asprone D, Durante M, Zinno A, Balsamo A, Prota A (2015) Structural behaviour of masonry panels strengthened with an innovative hemp fibre composite grid. Constr Build Mater 100:111–121
8. Olivito RS, Dubois F, Venneri A, Zuccarello FA (2012) Experimental and numerical analysis of masonry macroelements reinforced by natural-fibre-composite materials. In: Proceedings of 4th international conference on FRP composites in civil engineering, Rome/Italy, 13–15 Jun 2012
9. Babaeidarabad S, De Caso F, Nanni A (2014) URM walls strengthened with fabric-reinforced cementitious matrix composite subjected to diagonal compression. J Composit Constr 18(2):04013045
10. Marcari G, Basili M, Vestroni F (2017) Experimental investigation of tuff masonry panels reinforced with surface bonded basalt textile-reinforced mortar. Compos B Eng 108:131–142
11. Wang X, Lam CC, Iu VP (2018) Experimental investigation of in-plane shear behaviour of grey clay brick masonry panels strengthened with SRG. Eng Struct 162:84–96
12. Almeida JAPP, Pereira EB, Barros JAO (2015) Assessment of overlay masonry strengthening system under in-plane monotonic and cyclic loading using the diagonal tensile test. Constr Build Mater 94:851–865
13. Parisi F, Iovinella I, Balsamo A, Augenti N, Prota A (2013) In-plane behaviour of tuff masonry strengthened with inorganic matrix–grid composites. Compos B Eng 45(1):1657–1666
14. Gattesco N, Boem I, Dudine A (2015) Diagonal compression tests on masonry walls strengthened with a GFRP mesh reinforced mortar coating. Bull Earthq Eng 13(6):1703–1726
15. Ferretti F, Incerti A, Ferracuti B, Mazzotti C (2017) FRCM strengthened masonry panels: the role of mechanical anchorages and symmetric layouts. Key Eng Mater 747:334–341
16. Incerti A, Tilocca AR, Ferretti F, Mazzotti C (2018) Influence of masonry texture on the shear strength of FRCM reinforced panels. In: Proceedings of the 11th international conference on structural analysis of historical constructions, SAHC2018, 2018, Cusco, Peru

17. Del Zoppo M, Di Ludovico M, Balsamo A, Prota A (2019) Experimental in-plane shear capacity of clay brick masonry panels strengthened with FRCM and FRM composites. J Composit Constr 23(5):04019038
18. De Lorenzis L, Galati N, Ombres L (2004) In-plane shear strengthening of natural masonry walls with NSM CFRP strips and FRCM overlay. In: Proceedings of the fourth international seminar on structural analysis of historical constructions, Padua, Italy, pp 847–856
19. Prota A, Marcari G, Fabbrocino G, Manfredi G, Aldea C (2006) Experimental in-plane behavior of tuff masonry strengthened with cementitious matrix-grid composites. J Composit Constr 10(3):223–233
20. Faella C, Martinelli E, Nigro E, Paciello S (2010) Shear capacity of masonry walls externally strengthened by a cement-based composite material: an experimental campaign. Constr Build Mater 24(1):84–93
21. Balsamo A, Di Ludovico M, Prota A, Manfredi G (2012) Masonry walls strengthened with innovative composites. American Concrete Insitute, ACI special publication, 2(275):769–786
22. Del Zoppo M, Di Ludovico M, Balsamo A, Prota A (2019) In-plane shear capacity of tuff masonry walls with traditional and innovative Composite Reinforced Mortars (CRM). Constr Build Mater 210:289–300
23. Babaeidarabad S, Arboleda D, Loreto G, Nanni A (2014) Shear strengthening of unreinforced concrete masonry walls with fabric-reinforced-cementitious-matrix. Constr Build Mater 65:243–253
24. Ismail N, El-Maaddawy T, Khattak N, Najmal A (2018) In-plane shear strength improvement of hollow concrete masonry panels using a fabric-reinforced cementitious matrix. J Composit Constr 22(2):04018004
25. Ferraral G, Caggegi C, Gabor A, Martinelli E (2019) Experimental study on the adhesion of basalt textile reinforced mortars (TRM) to clay brick masonry: the influence of textile density. Fibers 7(12):103
26. Ferrara G, Caggegi C, Martinelli E, Gabor A (2020) Shear capacity of masonry walls externally strengthened using Flax-TRM composite systems: experimental tests and comparative assessment. Constr Build Mater 261:120490
27. Ferrara G, Caggegi C, Gabor A, Martinelli E (2019) Mechanical behaviour of masonry panels strengthened by Flax TRM systems. In: Key engineering materials, vol 817, pp 427–434

Chapter 6
Observations on the Displacement Field of Strengthened Walls Subjected to Diagonal Compression Test

As shown in the previous chapters, although plant-based TRM systems present a high potential, they are also characterised by several limitations that prevent their spread in the market. Most of the drawbacks are related to the variability of the properties of plant textiles, due to the nature of the fibres and to the fabric assembly techniques. A standardisation of textile production techniques, specifically conceived for applications in TRM composites, would help overcome some issues, such as the stiffening of the fibre bundles, or the thickening of the fabric. Many other aspects are still under investigation, such as coating and impregnation processes, fibres physical and chemical treatments, refinement of the geometry of the TRM composites.

These innovative techniques may lead to improvements in the mechanical performance of plant-based TRM composites. However, it is technically onerous to carry out experimental investigations on the structural scale of analysis, especially when several different solutions have to be compared. Therefore, the mechanical response is rather assessed on the composite scale of analysis, by means of textile and TRM tensile tests, pull-out tests to investigate the fibres-to-matrix bond, and shear bond tests to analyse the composite-to-substrate adherence.

Therefore, it is fundamental to define the parameters on which it is necessary to focus at the materials scale of analysis, to obtain an improvement of the response on the structural scale. It is necessary, to this purpose, to define a correspondence between the mechanical response of TRM systems tested at the composite scale, with the mechanical response of structural element externally strengthened by the TRM.

The aim of this chapter is to find a relationship between the strain field on external surface of Flax TRM strengthened walls tested in diagonal compression, and that of the external surface of Flax TRMs tested in tension. The objective is to define the crucial parameters on which it is necessary to focus during tensile test in order to have an improvement of the system when applied on structural elements.

G. Ferrara, *Flax-TRM Composite Systems for Strengthening of Masonry*, Springer Theses,
https://doi.org/10.1007/978-3-030-70273-1_6

6.1 Methods

The analysis was carried out by processing DIC images concerning the diagonal compression test of a representative specimen, i.e. the specimen SW-2L-1. A total of 26 images, representative of significant steps of the test, were processed. Six images represent the first phase of the response up to the peak load, the others are representative of the post peak phase. A picture, the number 26, was selected to represent the very ultimate state of the wall. Figure 6.1 shows the load-vertical displacement curve of the specimen SW-2L-1, on which the points corresponding to the selected photos are highlighted.

The value of the load, the time of the photo captured, and the vertical displacement corresponding to each image, are reported in Table 7.6.

The analysis provides the displacement field on the entire surface of the specimen. However, with the aim of extrapolating parameters to compare to the tensile test response of TRMs, in this analysis the displacements along the horizontal diagonal were taken into consideration. According to the hypothesis of pure shear conditions, such direction represents the principal stress direction in tension, therefore it is the direction that comes closest to the pure tensile conditions reproduced in tensile tests. The scheme of the images processed is shown in Fig. 6.2, where the zone and section of analysis, the reference system, the loaded corners and the two diagonals are highlighted.

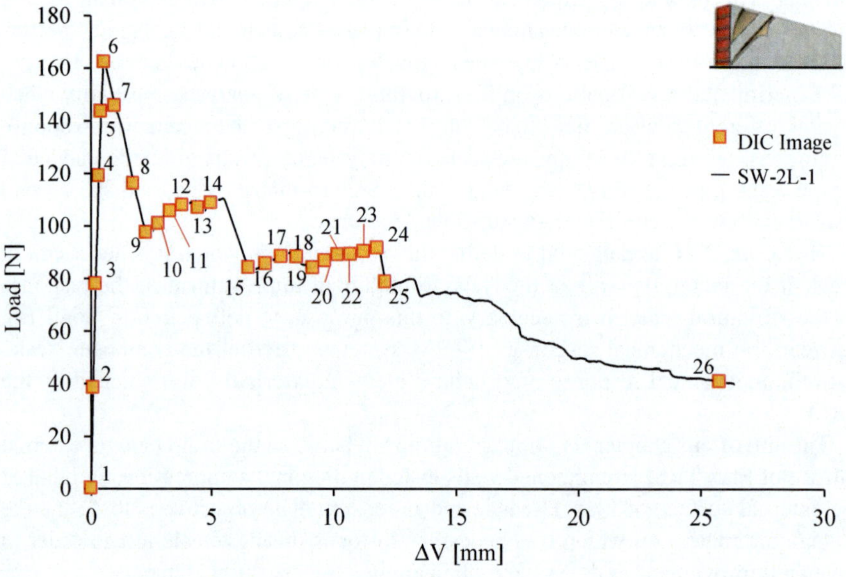

Fig. 6.1 Load-vertical displacement curve of specimen SW-2L-1 with the selected photos highlighted by yellow squares

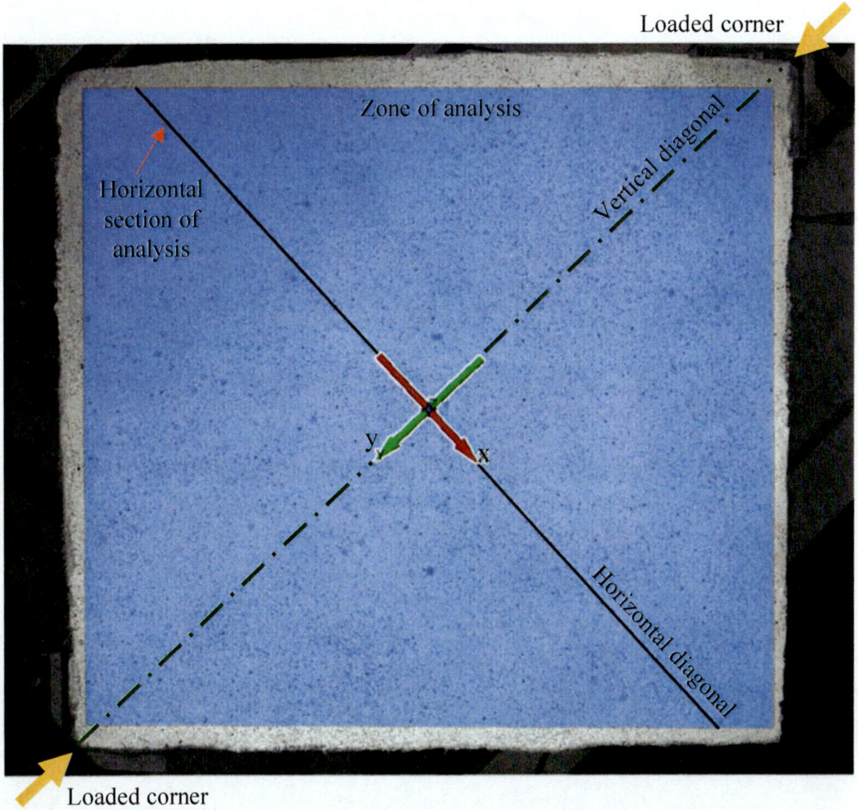

Fig. 6.2 Scheme of the images processed by means of DIC analysis

6.2 Analysis

From each image the value of the of the displacements along the x direction (herein after referred to as horizontal displacements) can be derived for all the points of the horizontal section of analysis (Fig. 6.3). When a crack appears on the wall surface, the displacement curve corresponding to that instant is characterised by a discontinuity in correspondence to the abscissa at which the crack occurred.

By estimating the difference of displacement between points placed at different edges of the crack, the crack opening can be assessed, and by monitoring such value for the other photos, the evolution of the crack openings can be observed.

From image 1 to 6 no discontinuities are shown being the curves referring to the elastic phase of the response up to the achievement of the peak load. From picture 7 to 15 several discontinuities can be observed, representing the development of several cracks in the post peak phase. From picture 15 on an increase of the cracks opening is observed without the occurrence of new cracks.

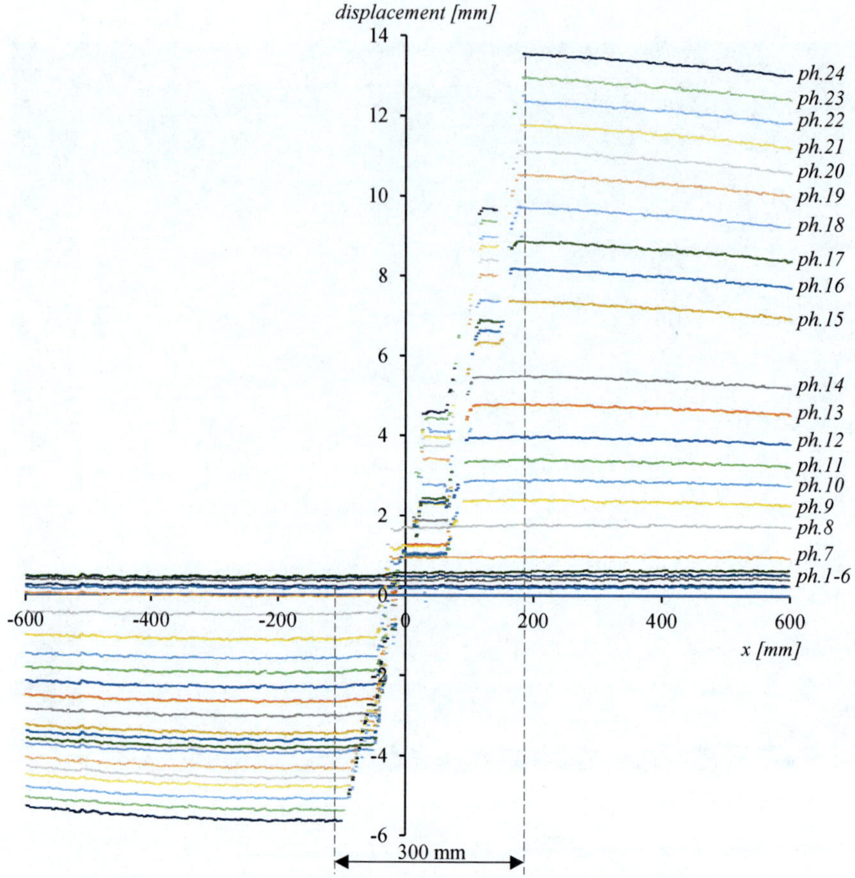

Fig. 6.3 Displacement along the horizontal analysed section for all the selected pictures

A further aspect that can be observed is that the displacements are concentrated in a zone of about 300 mm in the middle of the length of the horizontal section of analysis. Such area, referred to as the zone of deformation, corresponds to the centre of the panel, where the shear and normal stress attain their maximum values. The displacement curves of some marker images, corresponding to significant changing in the curve of displacement due to the occurrence of a crack, are shown more in detail with respect to the central zone of deformation analysed (Fig. 6.4).

The horizontal displacement field of the entire surface for the significant pictures, deriving from the DIC analysis, is shown in Fig. 6.5. Such representation emphasises the crack development and the concentration of the deformation in the central zone of the panel.

Image 26 (Fig. 6.5h), not considered in the analysis, was shown so as to represent the ultimate state configuration of the specimen. By analysing the displacement in the central deformation zone, it is possible to assess, for each image, a mean value of the

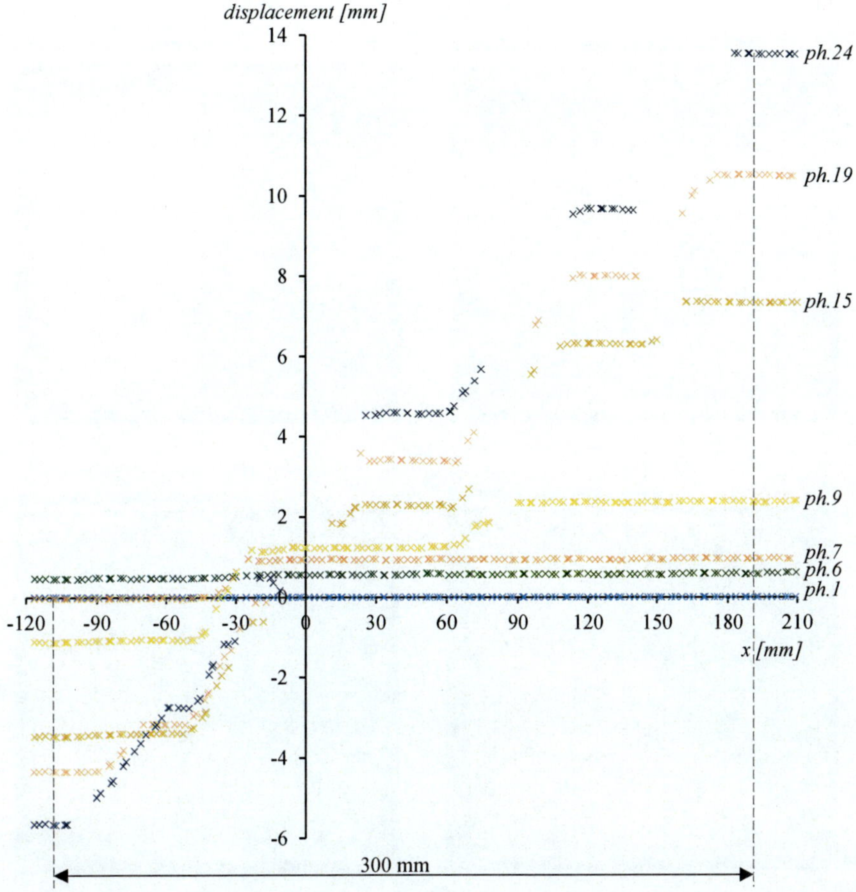

Fig. 6.4 Displacement along the zone of deformation for significant images

deformation, ε_{mean} (ratio between the difference of the displacements corresponding to the two edges of the zone of deformation and the length of this latter, 300 mm). It represents an approximated estimation of the strain of the most stressed point of the panel. The values of ε_{mean}, for each selected photo, are reported in Table 6.1. The values are also plotted in Fig. 6.6 with respect to the vertical displacement attained during the test.

The mean value of the strain of the deformation zone can be compared with the axial strain deriving from TRM tensile tests. It is worth highlighting that the boundary conditions of the composite applied to the masonry are not the same of the conditions of pure tension to which the TRM is subjected in tensile tests. Moreover, along the horizontal direction of the walls the textile is differently oriented with respect to the axial direction of tensile coupons, rotated of an angle of 45°. Therefore, it cannot be assumed that a real correspondence between the two different behaviours exists.

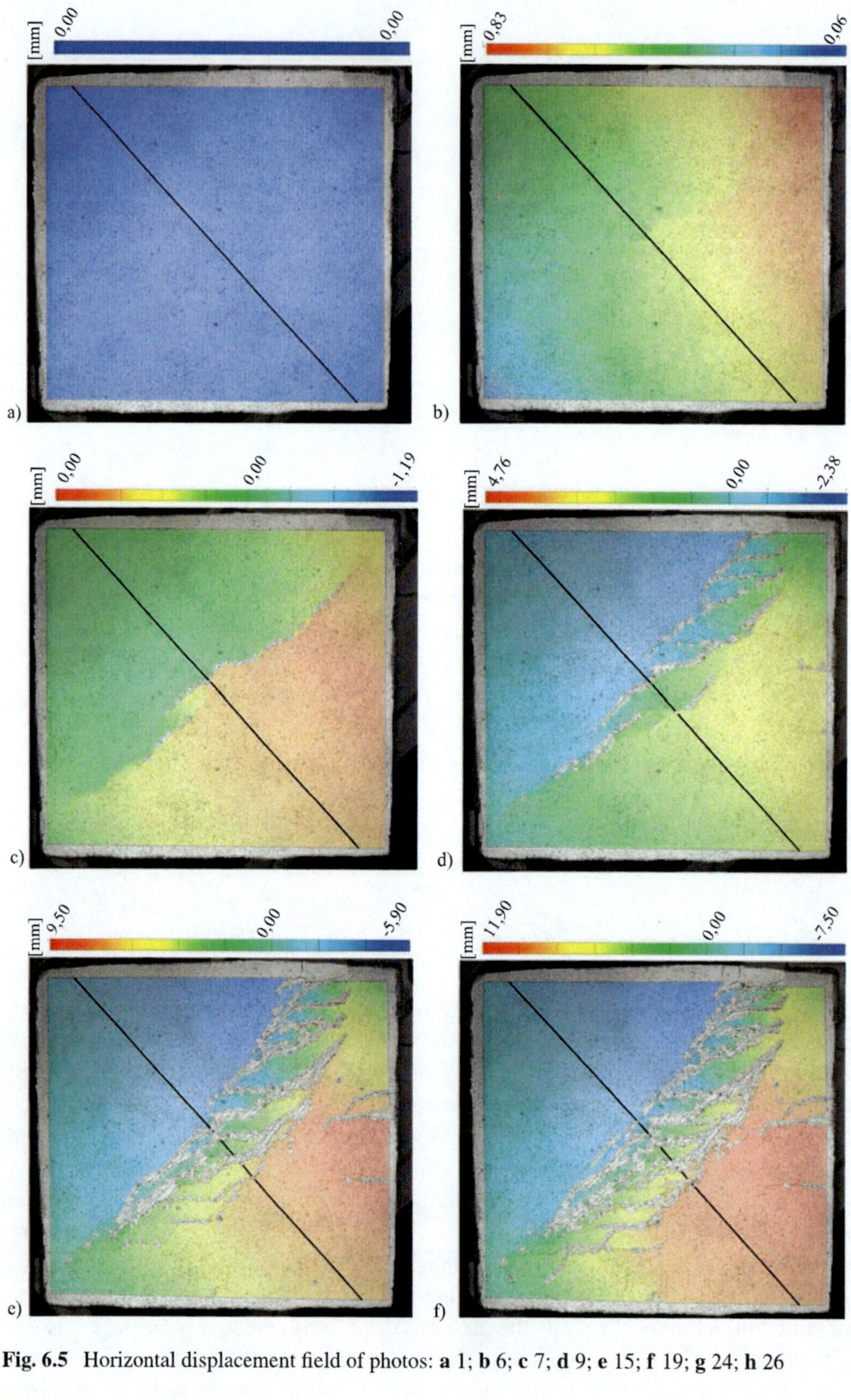

Fig. 6.5 Horizontal displacement field of photos: **a** 1; **b** 6; **c** 7; **d** 9; **e** 15; **f** 19; **g** 24; **h** 26

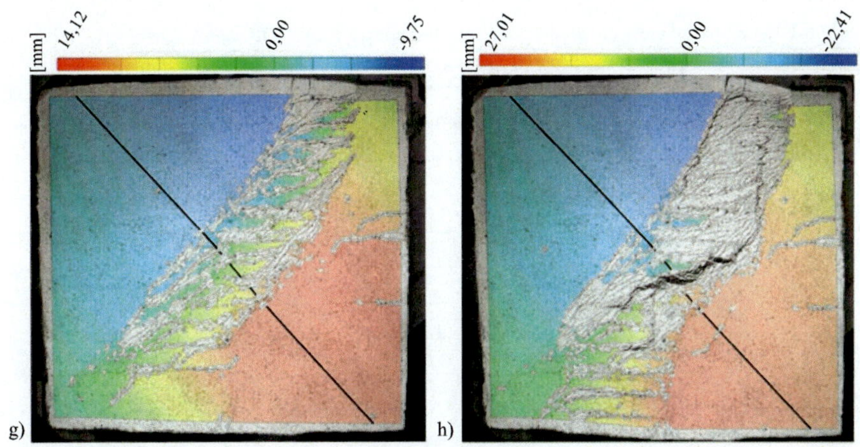

Fig. 6.5 (continued)

However, it may still be interesting to compare the strain field exhibited by the composite in tensile tests, with the strain field of the portion of the wall solicited in tension.

To this purpose, the tensile response of the specimen of the series Flax TRM-2L-T are plotted in Fig. 6.7. In the graph the mean values of the strain and stress corresponding to the transition between the different stages, are highlighted.

With respect to Table 6.1, it can be observed that in the range of photos 1-6, during the elastic phase of the diagonal compression response, the mean strain is included in the range of $0.00 \div 0.05\%$. Being such values less than the value ε_1 (the mean deformation of the transition point between Stages I and II in the tensile test response), it can be assumed that in correspondence of the elastic branch of diagonal compression response, the TRM works under elastic conditions. As a matter of fact, the TRM mechanically behaves as in the Stage I of the tensile response.

It is difficult to find an accurate correlation with respect to the wall post-elastic behaviour, due to the gradual loss of the pure shear conditions in the plastic phase.

However, it can be observed that in correspondence to image 15 a value of mean strain, ε_{mean}, is attained equal to 3.60%. Such is results very close to the mean value of the transition point between Stages II and III, ε_2. For the reasons explained above such similitude in the values does not represent an actual correspondence. However, it may be assumed that, in the post peak diagonal compression response, the TRM within the zone of deformation, exhibits a behaviour characterised by a range of strains similar to that exhibited during Stage II in tensile test conditions. Therefore, may be reasonable to associate the post elastic behaviour in diagonal compression, characterised by several drops of loads corresponding to cracks in the wall, to the response of the TRM in tension during Stage II.

It is also interesting to remark that in the range of photos 15–25 no development of new cracks is observed, and the existing cracks increase their opening. Moreover, within this range, the response is characterised by a slightly hardening behaviour.

Table 6.1 Main parameters concerning the selected photos for the DIC analysis

Photo	Time	Load	ΔV	ε_{mean}
	(s)	(kN)	(mm)	(%)
1	0	0	0.00	0.00
2	88	39	0.09	0.01
3	172	78	0.20	0.02
4	249	119	0.33	0.02
5	295	144	0.42	0.02
6	333	163	0.54	0.05
7	335	146	0.98	0.34
8	339	116	1.77	0.76
9	344	98	2.29	1.16
10	379	101	2.81	1.49
11	414	106	3.29	1.77
12	449	108	3.82	2.09
13	484	107	4.45	2.48
14	519	109	5.01	2.83
15	554	84	6.55	3.60
16	589	86	7.25	3.92
17	624	89	7.89	4.21
18	659	88	8.51	4.55
19	694	84	9.16	4.95
20	729	87	9.64	5.23
21	764	89	10.16	5.51
22	799	89	10.70	5.81
23	834	90	11.23	6.11
24	870	92	11.78	6.39
25	875	79	12.11	6.40
26	1692	41	25.73	12.19

Starting from the image 25 a gradual decrease of strength is observed due to the progressive rupture of the textile along the vertical diagonal. Such behaviour can be associated to the ultimate configuration of tensile tests corresponding to the failure of the fibres.

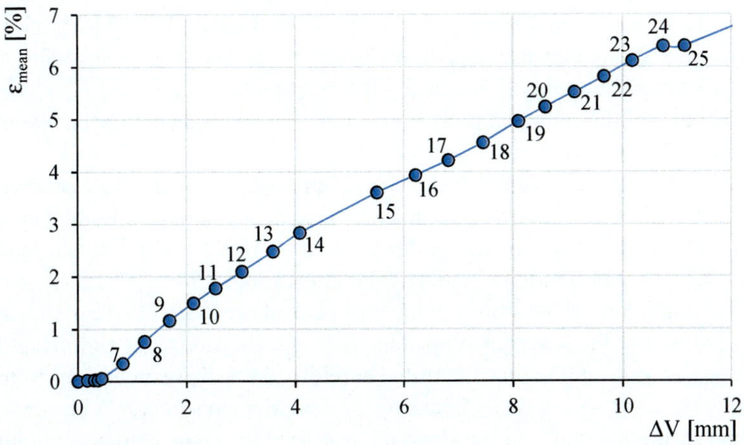

Fig. 6.6 Mean strain in the zone of deformation versus vertical displacement

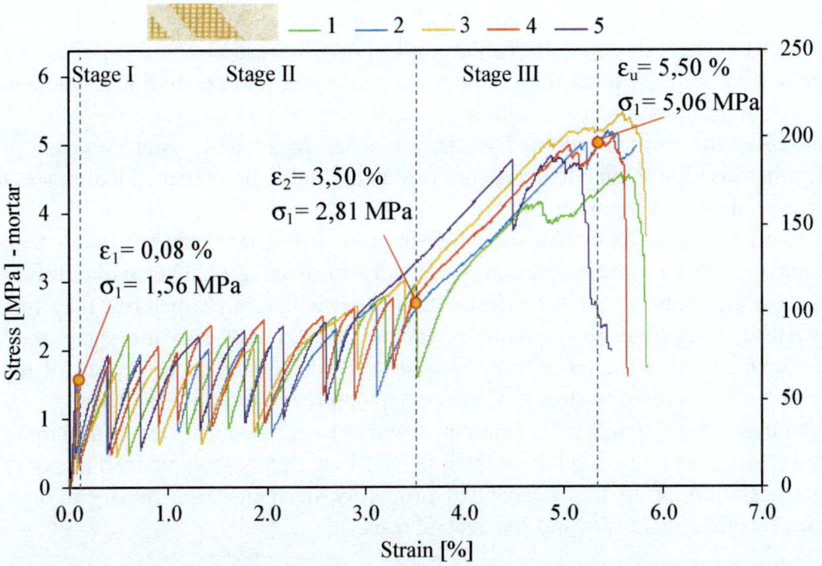

Fig. 6.7 Stress-strain response of specimens of the series Flax TRM-2L-T

6.3 Observations

With the aim of optimizing the global performance of the system, leaning towards a diagonal compression test response in which the post-elastic response comprises a hardening phase with an ultimate load higher than the elastic one, it is necessary to improve the tensile response of TRMs.

From the comparison between the tensile behaviour of Flax TRMs and their response once applied on the structural element, in terms of displacements, hence strains, it is possible to obtain interesting information.

On the basis of the considerations explained before, we can define the parameters on which the attention has to be focused on.

Having shown that the composite behaves in an elastic phase up to the achievement of the peak load, we can assume that a TRM system characterised by higher values of the normal stress σ_1 (transition point between Stages I and II) would lead to a higher value of the elastic load in diagonal compression.

The correlation between strains in the post peak phase, with the strain in Stage II, may lead to assume the post peak drop of load being associated to the significant drops of load observed during the tensile test. Therefore, a tensile response characterised by smaller values of the strain ε_2, hence by the developments of cracks closer to each other, and by higher values of the stress σ_2, may lead to elastic response in diagonal compression characterised by less significant losses of loads. Moreover, a higher stress level guaranteed by adequate textile amounts, proportionated with respect to the unreinforced wall strength, may even lead towards a hardening behaviour in the post elastic phase.

Several technical solutions are available to improve the response, such as stiffening of the textile, improving of the textile-to-mortar bond, matrix modifications, and optimisation of the reinforcement ratio.

Whatever the solution adopted is, the efficiency of a TRM system in view of applications as reinforcement of structural elements, has to be evaluated with respect to the considerations reported above.

It is worthy underlying that the analysis proposed in this chapter, rather than defining an accurate correspondence between the local and global behaviour, represents a starting point to further investigations, that with similar methods may find an exhaustive relationship, perhaps by taking into account also the shear-bond test behaviour, being the latter more representative to the actual behaviour of the composite once applied to structural elements, with respect to the tensile one.

A comprehensive model in these terms, would be fundamental in view of designing formulas aimed at assessing the strength of TRM externally strengthened masonry elements starting from the mechanical properties deriving from the local scale characterisation, i.e. tensile and shear-bond tests.

Chapter 7
Influence of Fibre Treatment and Matrix Mix Design Modification on the Behaviour of Flax TRMs

As shown in previous chapters, Flax TRMs are characterised by a promising mechanical behaviour. However, there is still room for improvements in order to design and manufacture a composite system being competitive, from a mechanical point of view, with respect to most common TRMs characterised by high strength industrial fibres [4]. In Chap. 2, within the analysis of the state of the art, it was shown that to face such issues and to improve the composite system mechanical response it is possible to apply fibre treatments mainly consisting in hornification [7], alkali treatments [11], coating procedures [9], matrix modifications, optimization of the textile mesh and of the reinforcement ratio. These treatments, specifically designed, may improve either the strength and/or the durability performance. Moreover, in Chap. 6 some considerations were reported in order to define the parameters to take into account to quantify the efficiency of an improvement technique, by means of TRMs tensile strength characterisation.

The present chapter aims at showing that by applying some of the improvement technique available in literature, it is possible to confer a better mechanical performance to the composite system [6]. Specifically, a fibre impregnation treatment, and a modification of the mortar matrix, were considered.

The aim of the study is to show that by adopting simple technique measures it is possible to improve the mechanical response of Flax TRMs. It is important to highlight that the experimental evidences showed in this chapter represent a preliminary step toward the definition of satisfying improvement techniques, rather than the proposal of a specific treatment.

The fibre treatment proposed consisted in the application of a latex-based substance on the flax textile. Prior to applying such treated textile as reinforcement in TRMs, the problem of the influence of the treatment on the fibre morphology and strength, and on the fibre-to-mortar bond behaviour, was addressed. Subsequently, the tensile behaviour of Flax TRMs, performed by means of the treated textile, was investigated in order to analyse the efficiency of the treatment. Finally, the tensile behaviour of a Flax TRM, performed by means of the treated textile and by adding

G. Ferrara, *Flax-TRM Composite Systems for Strengthening of Masonry*, Springer Theses,
https://doi.org/10.1007/978-3-030-70273-1_7

within the matrix short curaua fibres, was assess as well with the aim of defining the contribution provided by improving the mortar matrix.

The experimental activity was carried out at the Laboratory NUMATS–COPPE of the Federal University of Rio de Janeiro (Brazil).

7.1 Influence of the Fibre Treatment on the Morphology and Mechanical Behaviour of Flax Yarns Embedded in Hydraulic Lime Mortar

This section aims at investigating the effect of flax fibres treatment on their morphology, tensile strength and mortar-to-fibres adherence behaviour. To this purpose flax yarns, extracted from a textile and coated by means of a polymeric material, are subjected to tensile and pull-out tests and their response, compared with dry flax threads, is discussed taking into account the yarns geometry assessed by means of Scanning Electron Microscope (SEM) images analysis.

It represents a preliminary step to observe the influence of the treatment on the local behaviour at the fibre-to-matrix interface, before further analysis to be carried out at the composite scale by means of TRM tensile tests.

7.1.1 Materials and Methods

The flax fibre and the hydraulic lime mortar employed are those whose physical and mechanical main parameters are shown in Chaps. 3 and 4.

A Carboxylated Styrene Butadiene Rubber (XSBR) latex was used to coat the flax yarns under investigation. This copolymer is largely used in textile industry, both with synthetic and pkant fibres, to improve the bond with the matrices [8].

It is worth emphasising the need of analysing in detail the sustainability of the XSBR coating and the environmental impact of the entire system. However, as a first step, it is fundamental to investigate the efficiency of the coating system from a mechanical point of view. At a later time, durability investigations and life-safety assessment analysis are needed to have a comprehensive identification of the efficiency of the coating substance, and eventually define if more sustainable solutions need to be employed.

The coating procedure included the following steps: extraction of a flax thread from the textile (Fig. 7.1a); immersion of the thread in the polymer, in environmental temperature conditions (Fig. 7.1b); removal of the excess substance from the external surface of the thread (Fig. 7.1c); drying of the thread for 24 h at a controlled temperature of 38 ± 2 °C. In order to guarantee an elongated and regular shape, the threads were accurately arranged in a wood frame during the drying phase (Fig. 7.1d).

Fig. 7.1 Fibres treatment: **a** Flax thread; **b** Fibre impregnation; **c** Embedded thread; **d** Thread drying

High resolution SEM images, depicting the cross section of the flax fibres before and after the coating, were taken to analyse the influence of the treatment on the morphology of the threads. The morphology of dry fibres and the procedure to assess cross section and perimeter are discussed in Chap. 3. The images were characterised by a magnification of 80X. A total of 30 coated fibres were compared with 30 dry ones. Figure 7.2 shows the analysed images of a representative sample of dry and coated flax yarn. The images were manipulated to get the geometric parameters, with the same procedure shown in Chap. 3. The Fibre Intrinsic Efficiency Ratio ($FIER_r$) was introduced as parameter to describe the irregular geometric shape of the threads. It is defined as the ratio between the actual yarn perimeter, P_f, and un equivalent perimeter, P_{eq}, representing the length of the circumference of the circle having the same area of the yarn [8].

Tensile strength was assessed on both the impregnated and non-impregnated flax yarns. The following series of specimens were considered:

- non-impregnated-T-tensile: the specimens are the flax threads representing the main element of the fabric (labelled as Flax T in Chap. 3). The samples had a length of 70 mm and were randomly extracted from the fabric (Fig. 7.3a);
- impregnated-T-tensile: specimens obtained by coating samples having the same features of non-impregnated-T-tensile threads (Fig. 7.3b).

The series included by 13 to 15 samples. The specimens had a gauge length of 50 mm and were clamped for a length of 10 mm in both the edges. Tensile tests were

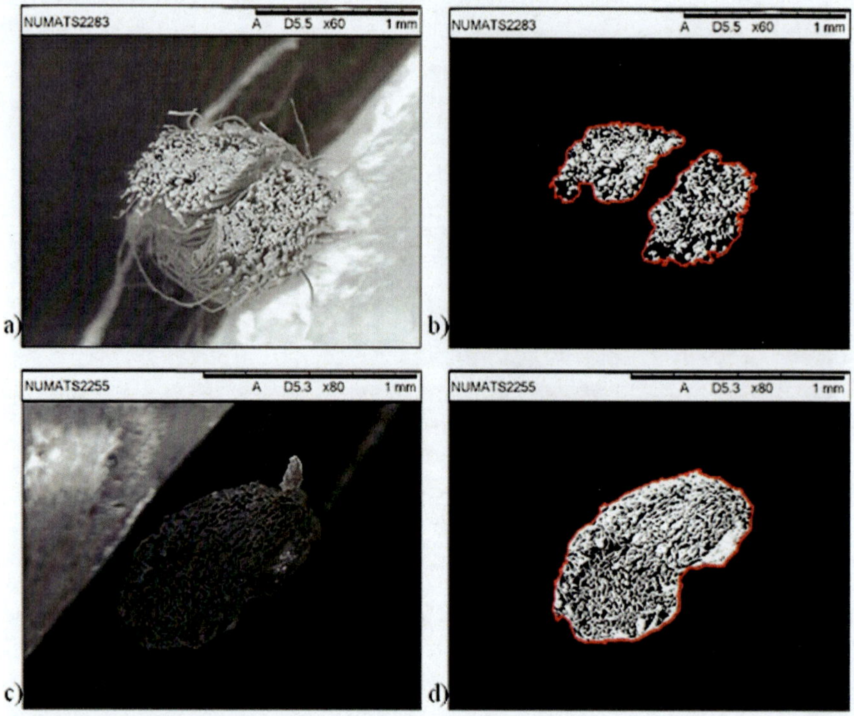

Fig. 7.2 Morphology of flax thread: **a** Non-impregnated thread SEM image; **b** Non-impregnates thread manipulated image; **c** Impregnates thread SEM image; **d** Impregnates thread manipulated image

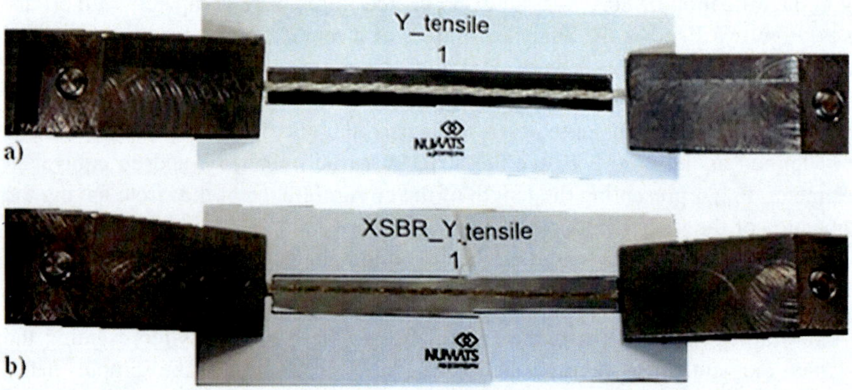

Fig. 7.3 Flax threads' tensile test: **a** Non-impregnated-T-tensile; **b** Impregnated-T-tensile

carried out in displacement control with an advancement rate of 4 mm/min, via a micro force machine Tytron 250 equipped with a 500 N load cell.

The bond between the flax threads and the mortar in which they were embedded was investigated via pull-out tests. To this purpose, cylindric hydraulic lime mortar specimens were cast with a diameter of 25 mm and a height of 25 mm, representing the thread embedded length. During the cast, a flax thread was placed along the axis of the small cylinder. The embedded length was defined so to have a slipping of the thread within the mortar, avoiding the tensile failure of the flax fibres. The following series of specimens were considered:

- non-impregnated-T-pullout: the specimens included uncoated flax threads (of the type non-impregnated-T-tensile) embedded in the mortar cylinder it consists of uncoated flax yarns embedded in the mortar;
- impregnated-T-pullout: the specimens included treated flax threads (of the type impregnated-T-tensile) embedded in the mortar cylinder it consists of uncoated flax yarns embedded in the mortar cylinders.

Both the series included 15 specimens. During the casting the threads, with a length of 35 mm, were immersed in the middle of the fresh mortar cylinder along its axis and embedded for the entire length of the cylinder. The specimens were tested after 28 days of curing in environmental conditions. Pull-out tests were carried out in displacement control with a rate of 1 mm/min via a micro force machine Tytron 250 equipped with a 500 N load cell. A cylindric steel device clamped the mortar element during the test. On the other side the free length of the flax thread was clamped over a length of 10 mm, and in proximity of the mortar. The clampers were placed at the edge of the mortar cylinder to make negligible the elastic elongation of the thread during the test (Fig. 7.4).

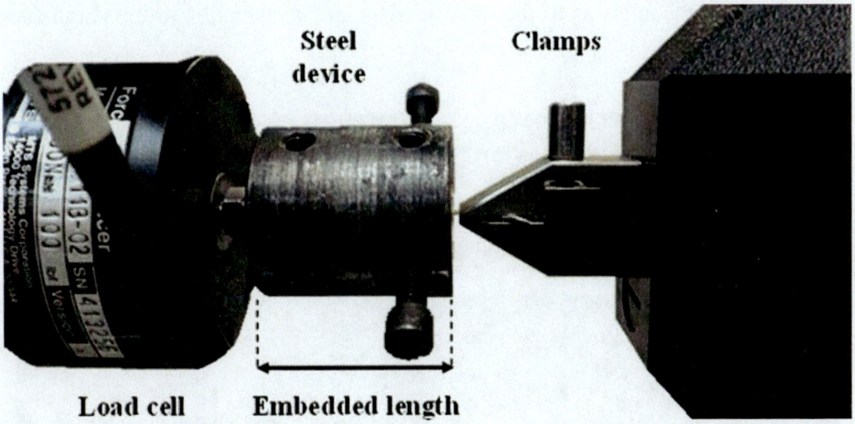

Fig. 7.4 Set-up of the pull-out test

7.1.2 Results and Discussion

7.1.2.1 Morphological Characterisation

A number of 30 images depicting non-impregnated threads and 30 images depicting impregnated threads were analysed to characterise the morphology of the fibres. SEM images of representative non-impregnated and impregnated specimens are shown in Fig. 7.5.

Both the series of specimens exhibited the same mean value of the cross section equal to 0.28 mm^2. Both the groups of threads, non-impregnated and impregnated, were extracted from the same portion of textile. Therefore, it is possible to assume that the coating treatment did not lead to any significant change in the cross-section area. In fact, the coating substance only coats the external part of the threads, without penetrating among the inner filaments of the yarn. This aspect appears clear in Fig. 7.6, where a sample of an impregnated yarn is depicted in different magnification scales.

As regard the perimeter of the flax threads, the two groups of fibres showed different values: the mean value exhibited by the non-impregnated specimen was 4.89 mm, while the mean value for the impregnated threads was 3.15 mm.

In addition to a similar mean value of the cross section, the two groups of specimens also show comparable frequency distribution of the values with no significant discrepancy (Fig. 7.7). Conversely, with respect to the perimeter, in addition to a different average value, the two groups of data follow different probabilistic distributions (Fig. 7.8).

The distribution of the relative frequency of the perimeter concerning the non-impregnated specimens assumes a "multi-modal" form, while it become "quasi uni-modal" for the impregnated specimens. As matter of fact, the impregnation reduces the heterogeneity of the geometry of the threads. The same evidence can be derived from the analysis of the relative frequency distribution of the parameter $FIER_r$ (Fig. 7.9).

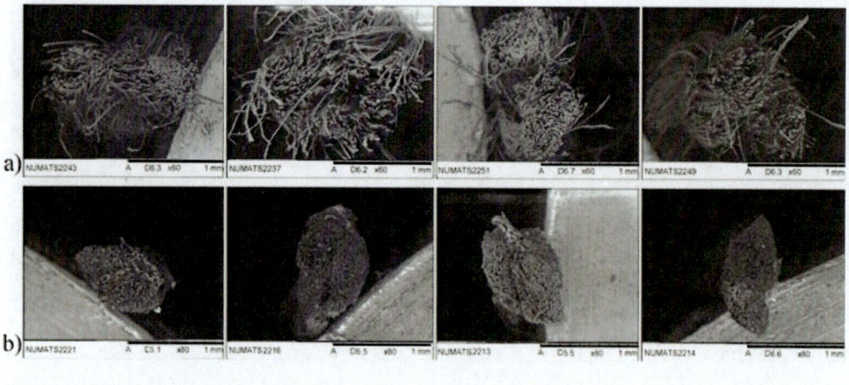

Fig. 7.5 SEM images of representative specimens of the series: **a** Non-impregnated-T-pullout; **b** Impregnated-T-pullout

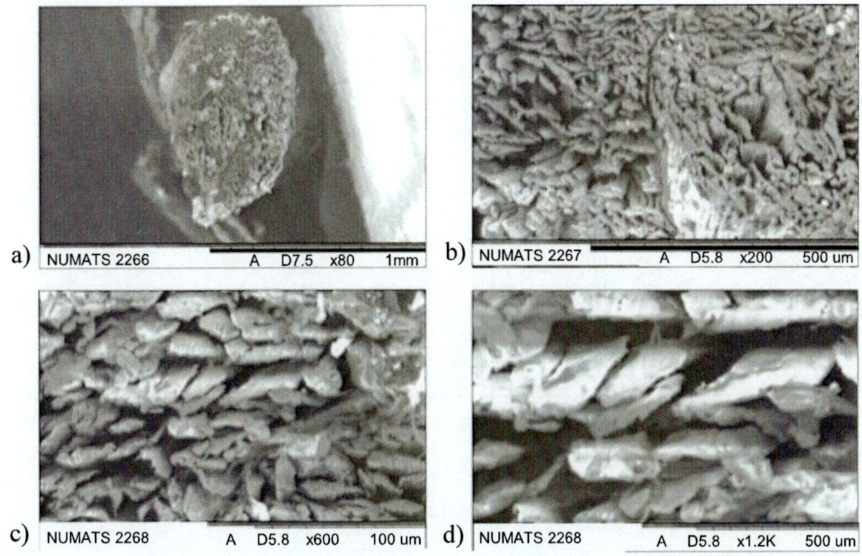

Fig. 7.6 SEM images of an impregnated yarn in different magnification scales: 80X, 200X, 600X, 1200X

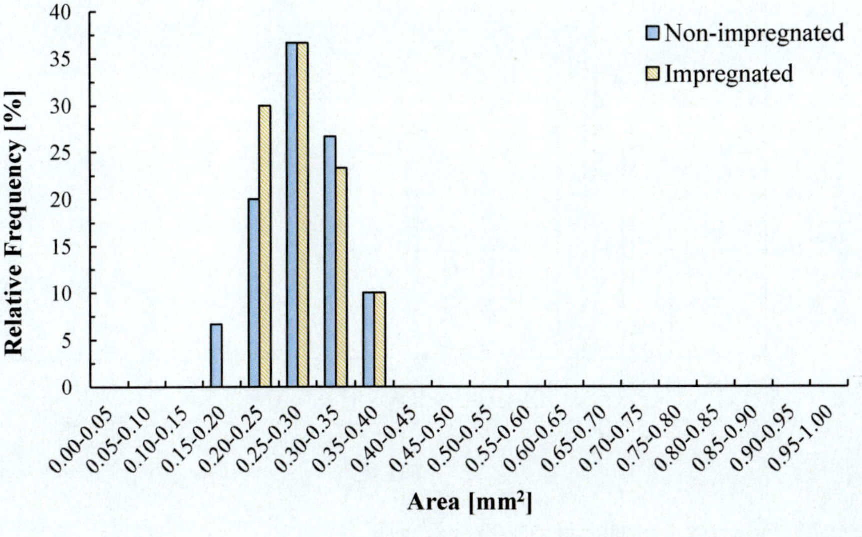

Fig. 7.7 Frequency distribution of the cross-section area

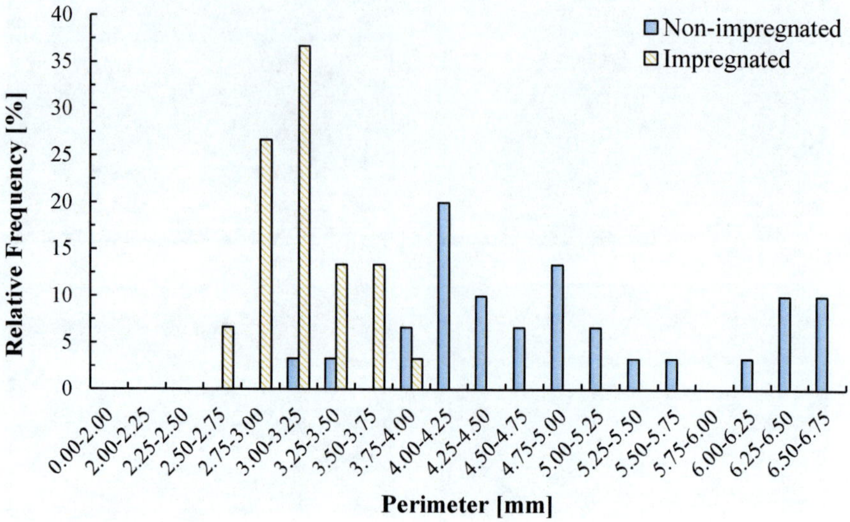

Fig. 7.8 Frequency distribution of the perimeter

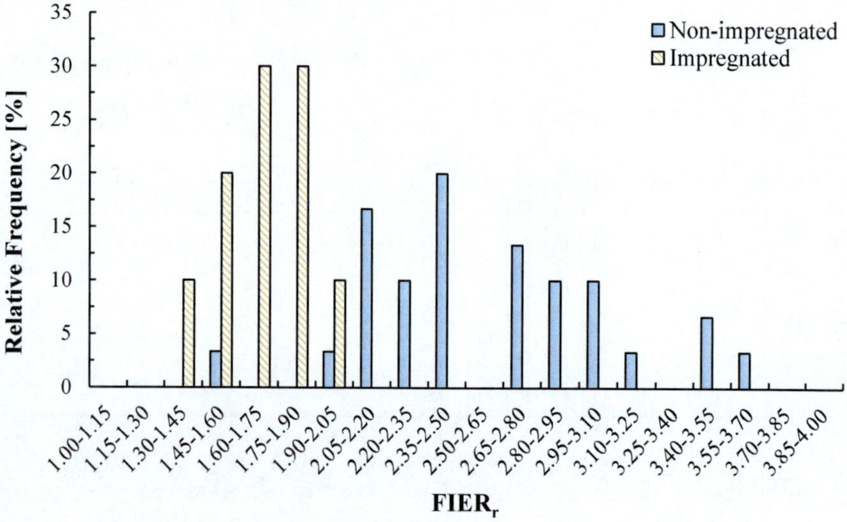

Fig. 7.9 Frequency distribution of the FIER$_r$ parameter

After the coating implementation the FIER$_r$ decreases of about the 35%, moving from a mean value of 2.61, for non-impregnated specimens, to a mean value of 1.69, for the impregnated specimens. Also in terms of dispersion of this parameter an influence of the coating treatment is shown, with a squeezed distribution for the impregnated threads and a more stretched distribution for the non-impregnated specimens Fig. 7.9. In fact, the morphology of the flax threads is significantly affected by the impregnation treatment that make them more regular in shape reducing the lateral surface of the bundles, due to a narrower space among the filaments, and diminishing the variability of the properties.

7.1.2.2 Tensile Behaviour

Axial displacement and applied load were recorded during tensile tests. The tensile behaviour was described by the follow parameters: the maximum force attained, F_{max}; the corresponding displacement, $disp_{max}$ and strain ε_{max}; the tensile strength, f_t; the Young's modulus, E, evaluated in the linear branch of the stress-strain curve, within the stress range from 50 to 90% of the tensile strength.

Table 7.1 list the mean values, with the respective coefficient of variations, of the mentioned parameters for both the non-impregnated-T-tensile and impregnated-T-tensile series of specimens. The stress-strain tensile response for both the series of specimens is shown in Fig. 7.10.

The stress-strain responses of both series of specimens included a lower initial stiffness increasing until the attainment of a constant value before the failure. This result is in line with what observed in Chap. 3, and, as discussed in the state of the art review, it is due to the particular microstructure of plant fibres, and to the macrostructure of the yarn, hence the arrangement of the filaments creating the threads. The impregnated-T-tensile specimens exhibited a mean value of the maximum strain, ε_{max}, significantly lower than that observed in the non-impregnated-T-tensile specimens, even if no significant differences in the Young's modulus were observed. This aspect may be attributed to the fact that the impregnated threads behaved as a "polymeric composite" strongly affected by the polymeric phase that also governs the failure mode. Non-impregnated specimens show a lower initial stiffness due to a stretching of the filaments of the thread when the load is applied. Conversely, due to the impregnation, this effect is less significant in the impregnated threads presenting a more "rigid" response in which a quasi-constant value of the stiffness is shown from the beginning of the test. In fact, the impregnation treatment created a macrostructure of the thread in which the initial response is stiffer. As result, the impregnated threads were less deformable. This lower deformability is a fundamental aspect in view of application of the flax textile as reinforcement in mortar based composite

Table 7.1 Mean mechanical properties of specimens tested in tension

Series	Number of specimens	F_{max}	$disp_{max}$	f_t	ε_{max}	E
		(N)	(mm)	(MPa)	(%)	(GPa)
Non-impregnated-T-tensile	15	100	3.0	341.2	6.1	9.06
		(9)	(10)	(20)	(10)	(19)
Impregnated-T-tensile	13	79	1.8	295.8	3.6	9.11
		(15)	(18)	(18)	(18)	(18)

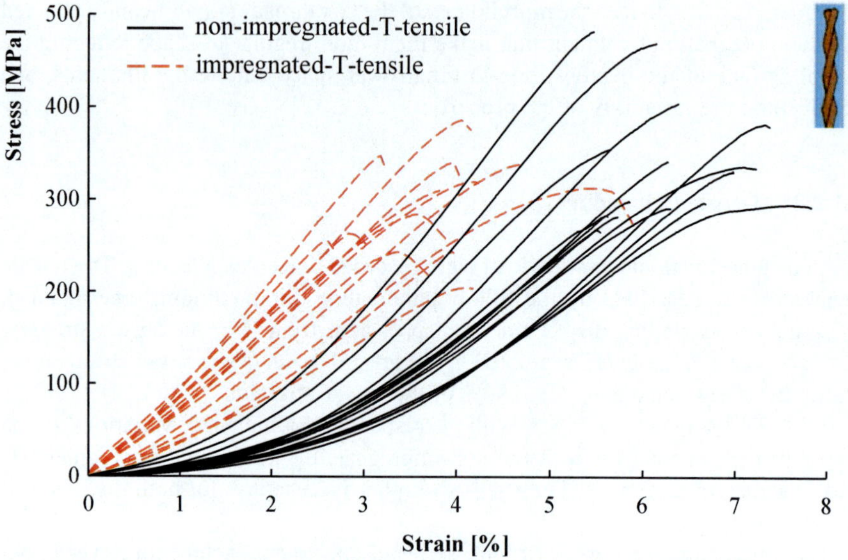

Fig. 7.10 Tensile test results of non-impregnated-T-tensile and impregnated-T-tensile series

systems. In fact, as discussed in previous chapters and emphasised by the literature [1], the significant difference in terms of deformability between mortar and textile represents one of the main limitations of these composite systems [5].

7.1.2.3 Bond Behaviour

Figure 7.11 shows the load-displacement curves concerning the pull-put tests of the npn-impregnated-T-pullout and impregnated-T-pullout series of specimens.

All the specimens exhibited a failure mode including the slipping of the flax thread within the matrix until the attainment of the complete pull-out of the fibres from the mortar cylinder. The load-displacement curves are in line with similar studies in which this debonding phenomenon occurred [8]. This response includes a bond adhesive phase, consisting in the first elastic branch, a drop of the load after the peak, and a frictional bond phase characterised by a quasi-constant value guaranteed by the residual frictional force. Some specimens exhibited, in addition to the first elastic branch, a mechanical bond phase, consisting in a non-elastic hardening branch until the achievement of the peak load. Most of the non-impregnated samples exhibited a mechanical bond phase, while the impregnated threads mainly had an elastic behaviour until the achievement of the peak load (Fig. 7.11).

The non-impregnated threads exhibited a softening behaviour in post peak phase, in which, due to a mechanical bond component, the sample could bear a load until the complete ejection of the thread from the mortar. The impregnated threads included a more pronounced softening phase with a residual frictional component that in most

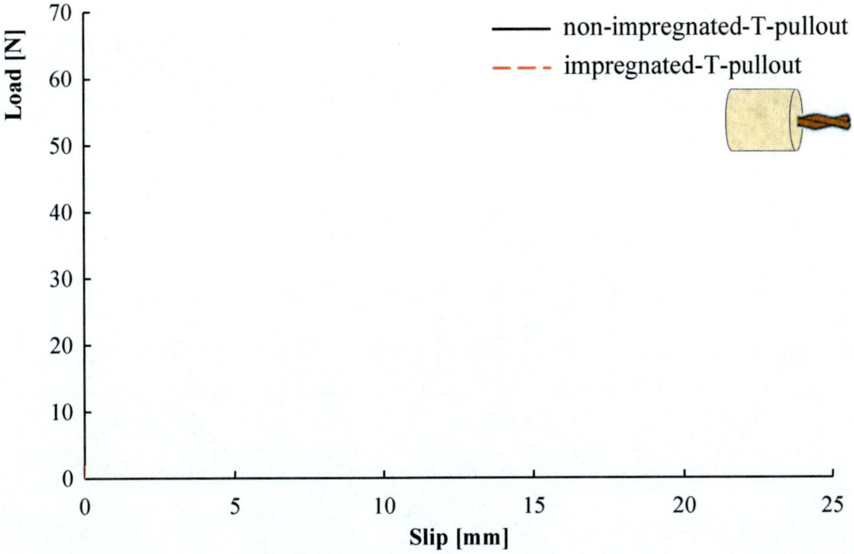

Fig. 7.11 Pull-out test results of non-impregnated-T-pullout and impregnated-T-pullout series of specimens

of the cases became null before the whole ejection of the thread from the mortar cylinder. The maximum load attained in the non-impregnated series of specimen, equal to 40.96 N, is two times higher than the value exhibited by the impregnated series of specimens, equal to 20.23 N. In fact, the impregnation procedure led to a reduction of the bond capacity between the threads and the matrix. This experimental evidence is in part depending on a loss of adherence between fibres and matrix, due to the presence of the coating substance. However, it is also in part attributed to the different morphology of the impregnated threads that, as shown above, had a smaller fibre-to-mortar interface surface.

The results are also presented in terms of shear stress at the fibre-to-mortar interface, τ_{max}, to analyse the bond properties regardless of the geometry of the samples. The shear stress was assessed by dividing the maximum force for the lateral surface of the threads (for each sample its own value of the perimeter was considered). Although showing a similar dispersion with respect to the relative frequency distribution (Fig. 7.12), the two series of specimens exhibited different main values of the shear stress, being the latter 0.37 MPa and 0.25 MPa respectively for the non-impregnated-T-pullout and the impregnated-T-pullout specimens. This discrepancy, consisting in a reduction of about the 30% of the maximum shear stress due to the use of the impregnation treatment, is less significant with respect to the difference highlighted by the comparison in terms of maximum loads. In fact, the impregnation provokes a reduction of the tortuousness of the threads. As matter of fact, the presence of the coating layer reduces the bond capacity of the system with respect to the case in which the flax threads are directly embedded within the mortar. This experimental

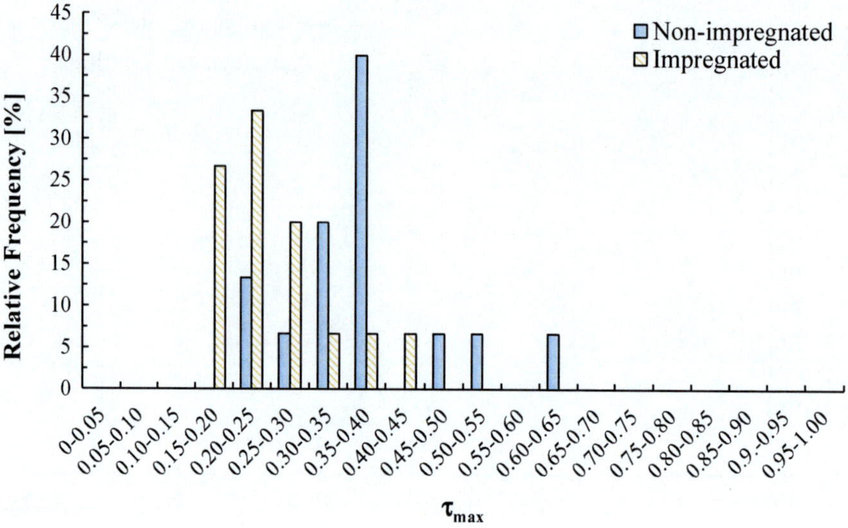

Fig. 7.12 Frequency distribution of shear stress of non-impregnated-T-pullout and impregnated-T-pullout series of specimens

evidence may be due to the nature of the polymer material adopted or to the type of coating treatment adopted. The impregnation created a weak layer between the threads and the matrix., resulting in a lower adhesion at the reinforcement-to-mortar interface surface.

The study represents a preliminary analysis for investigate the efficiency of the coating treatment adopted. It highlights the need of further studies aimed at improving the coating treatment with a beneficial effect not only in terms of stiffness of the fibres, but also in terms of fibre-to-matrix bond capacity. In addition, since the degradation of plant fibres in the alkaline environment of cement and lime based mortars is one of the main concerns hindering their use, it is necessary to investigate as well the influence of coating treatment on the durability of the composite system. Therefore, a more comprehensive analysis embracing all these fundamental aspects would be extremely importance to promote the use of sustainable retrofitting solutions comprising plant based TRM systems.

7.2 Influence of Fibre Treatment and Matrix Mix Design Modification on the Tensile Behaviour of Flax TRMs

As shown, the fibre treatment by means of a coating application affects the morphology and the tensile behaviour of flax threads, and also influences the bond between fibres and mortar.

With the aim of investigating the influence of such treatment on the mechanical behaviour of the composite, tensile tests are proposed on Flax TRMs performed by adopting textiles treated by means of the same coating treatment showed above. In addition, with the aim of achieving a more improved performance, the influence on the mechanical behaviour of matrix modifications, obtained by adding short fibres, was investigated as well.

7.2.1 Materials and Methods

The textile adopted to perform the TRM coupons consisted of the flax fabric whose mechanical and physical properties were discussed in Chap. 3. The treatment applied on the textile consisted in the application of a Carboxylated Styrene Butadine Rubber (XSBR) latex on the surface of the fabric in order to reproduce the same treatment performed on the flax thread subjected to tensile and bond characterisation and whose behaviour is discussed in the previous paragraph.

The treatment was carried out on flax fabric portions 35 cm large and 60 cm long. The textile portions were arranged on a wood frame specifically conceived to keep them strained during the entire coating procedure. The coating substance was applied on the textile in environmental temperature conditions (Fig. 7.13a), then the excess liquid was removed and the textile portions were dried at a controlled temperature of 38 ± 2 °C for 24 h (Fig. 7.13c). After the treatment flax strips, 60 mm large, were cut

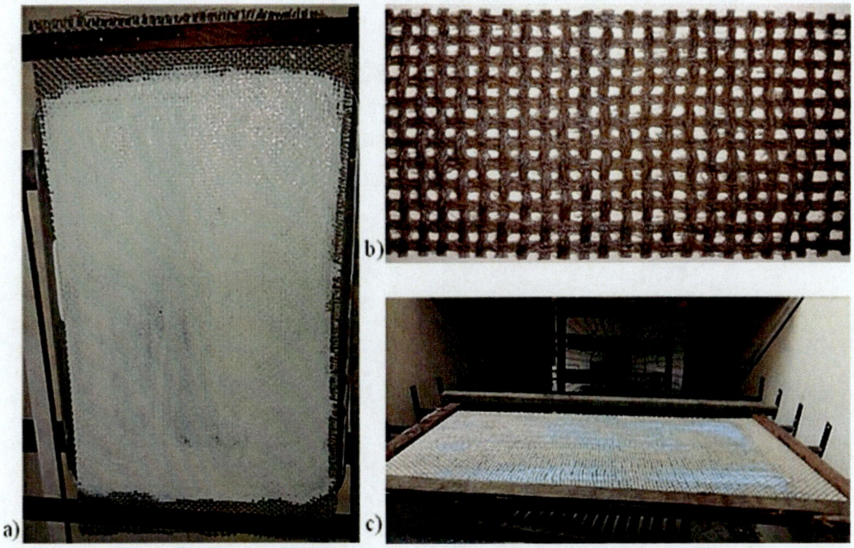

Fig. 7.13 Textile impregnation procedure: **a** Application of the coating substance; **b** Coated flax strip; **c** Drying of the textile

from the textile to be used as reinforcement in TRMs (Fig. 7.13b). The mechanical characterisation of the strips was carried out by means of tensile tests, considering the follow series of specimens:

- Flax fabric 6 cm, consisting in 5 specimens of flax textile with a width of 60 mm (containing 12 Flax-T threads) and a length of 500 mm;
- Flax fabric 6 cm-imp, consisting in 5 specimens of impregnated flax textile with width of 60 mm (containing 12 Flax-T threads) and length of 500 mm.

Tensile tests were carried out by means of a Shimadzu (model AG-Xplus) universal testing machine with 10 kN load capacity, with a gauge length of 300 mm, in displacement control with a rate of 4 mm/min. Each edge of the flax strips was glued by means of a resin within two aluminium plates for a length of 100 mm. The elongation of the textile during the test was taken via extensometers placed at mid-length recording the relative displacements over a length of 200 mm (Fig. 7.14).

Flax TRM specimens to be tested in tension were performed with the same manufacturing process already shown in Chap. 4, by alternating layers of mortar and textile (Fig. 7.15a, b). Specifically, the specimens were characterised by the TRM-2L reinforcement configuration in which two layers of textile, alternated by a thin layer of

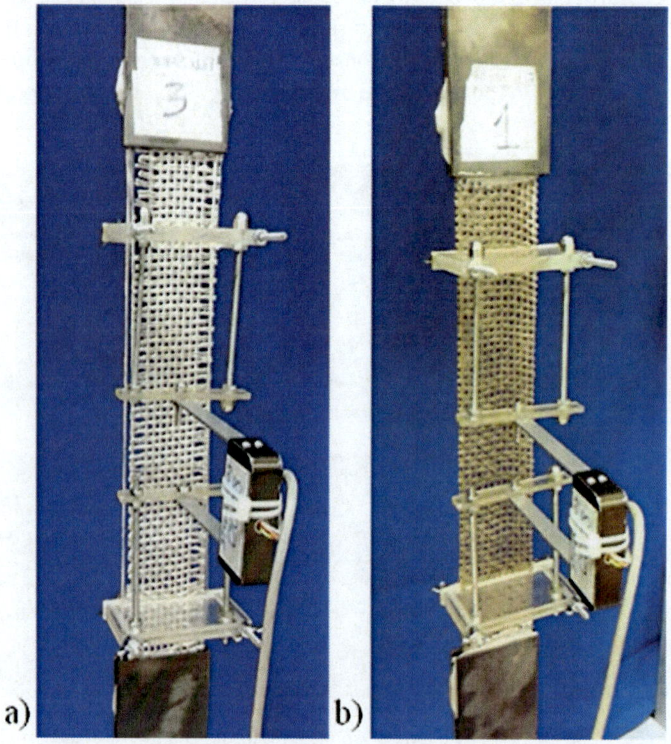

Fig. 7.14 Tensile test set up of **a** Flax fabric 6 cm and **b** Flax fabric 6 cm-imp specimens

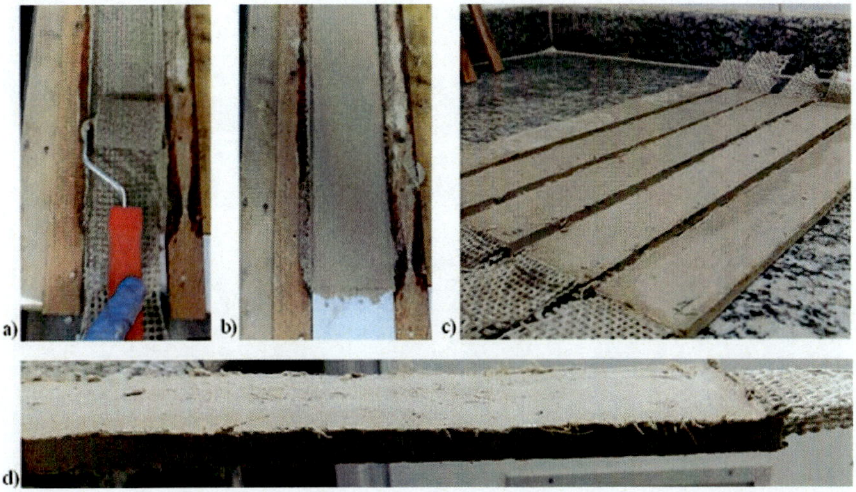

Fig. 7.15 **a** Application of the impregnated textile within the mortar; **b** Smoothing of the outer layer of mortar; **c** Flax TRM-2L-imp specimens; **d** Representative Flax TRM-2L-imp-SF specimen

mortar, were arranged within the sample. The series of specimens considered are shown as follows:

- Flax TRM-2L, consists of a series of specimens characterised by two layers of non-impregnated flax textile;
- Flax TRM-2L-imp, consist of a series of specimens characterised by two layers of impregnated flax textile (Fig. 7.15c);
- Flax TRM-2L-imp-SF, consist of a series of specimens characterised by two layers of impregnated flax textile, and by a short-fibre reinforced mortar (Fig. 7.15d).

For each series of specimens the mechanical properties of the mortar were assessed by means of bending and compression tests carried out according to specific standards [3]. The consistence of the fresh mortar properties was assessed as well by means of flow table test, and was expressed in terms of mean diameter of the mortar after jolting the table for 15 times [2].

The mix design adopted for the mortar is the same shown in Chap. 4. With respect to the series Flax TRM-2L-imp-SF, short fibres were added to the mix design. In accordance to the principles of sustainability on which the entire research study is based, plant fibres were chosen to strengthen the mortar. Specifically, 20 mm long curaua fibres were added to the mix design in a percentage of 1% in weight of the fresh mortar (Fig. 7.16a). The fibres had a diameter of 77.7 μm [10]. Before their use the fibres were subjected to chemical and physical treatments to improve their strength and their durability once immersed in alkaline mortars [11]. As first they were submitted to three hornification cycles, each one consisting in 3 h of immersion in 80 °C and 24 h of drying at 40 °C in a ventilated chamber. Consequently, an alkaline treatment was adopted consisting in 50 min of immersion in an alkaline

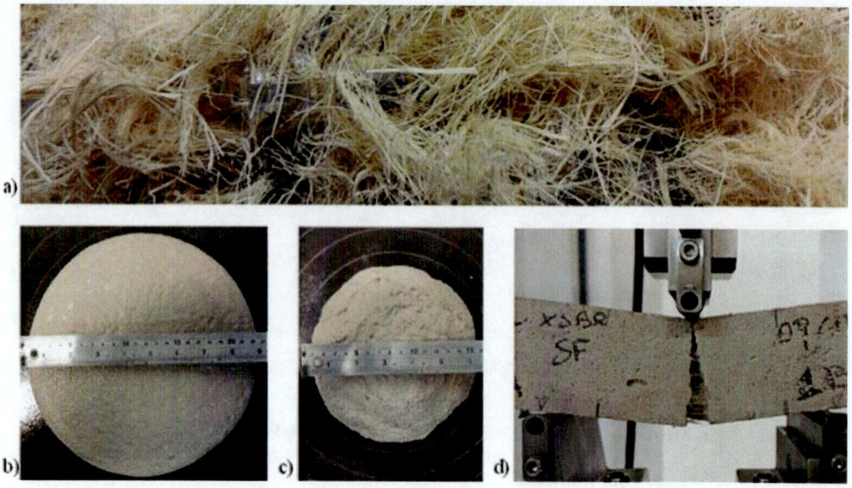

Fig. 7.16 **a** Curaua short fibres; **b** Consistence of fresh mortar (Flax TRM-2L-imp series); **c** Consistence of fresh mortar (Flax TRM-2L-imp-SF); **d** Bending test of curaua short fibre reinforced mortar (Flax TRM-2L-imp-SF series)

solution characterised by a concentration of 0.73% g/ml of $Ca(OH)_2$, followed by 24 h of drying at 40 °C in a ventilated chamber.

The addition of the short fibres within the mortar significantly increased the consistency of the fresh mortar (Fig. 7.16b, c). However, the short fibres conferred to the mortar a higher flexural strength and a post peak behaviour in flexion, thanks to their capacity to transfer the load between two edges of a crack within the mortar (Fig. 7.16d).

The main properties of the three series of specimens are listed in Table 7.2. All specimens were characterised by a length of 500 mm and a gauge length of 300 mm. Each edge was glued between two aluminium plates by means of a resin for a length of 100 mm. The gripping configuration was the clevis type. The test set up was arranged so that, by means of hinges, no bending and torsional moments were applied to the specimen during the test. In order to be consistent with the experimental evidences

Table 7.2 Main properties of the tensile test series of specimens

Series	Number of spec.	Section (mmxmm)	$A_{textile}$ (mm^2)	Fibre vol. ratio (%)	$f_{textile}$ (MPa)	$f_{c,mortar}$ (MPa)	$f_{t,mortar}$ (MPa)	$d_{flow\ table}$ (mm)
Flax TRM-2L	5	8 × 60	12.2	2.55	331	10.4	4.4	238
Flax TRM-2L-imp	5	8 × 60	12.2	2.55	266	8.4	3.9	223
Flax TRM-2L-imp-SF	5	8 × 60	12.2	2.55	266	9.9	5.9	153

Fig. 7.17 **a** Tensile test set up; **b** Light spots and camera and **c** Surface treatment for DIC

showed in the previous chapters, the test set up was conceived to be as similar as possible to the one showed in Chap. 4. However, being the experimental analysis carried out at a different laboratory, the occurrence of differences in the specimens manufacturing process, and in the testing, was unavoidable. For this reason, the reference series, i.e. the Flax TRM-2L, although having the same characteristics of the series Flax TRM-2L-T showed in Chap. 4, was implemented again.

Tensile tests were carried out by means of a Shimadzu (model AG-Xplus) universal testing machine with 10 kN load capacity, in displacement control with a rate of 0.3 mm/min (Fig. 7.17a, b).

The external surface of the specimens was treated, by creating a white background on which black dots were stochastically sprinkled, in order to get images for Digital Image Correlation (DIC) analysis (Fig. 7.17c). The acquisition of the photos was carried out by means of a camera placed at a focal distance of 50 cm. The acquisition rate was of 3.75 ph/min and the resolution of the photos was of 4310 × 2868 pixels (0.08 mm/pix).

7.2.2 Results and Discussion

The tensile response of the impregnated and non-impregnated flax textile is shown in terms of stress-strain in Fig. 7.18.

The main parameters are listed in Table 7.3. As expected, in accordance with what was observed in previous chapters, the tensile response was characterised by a first

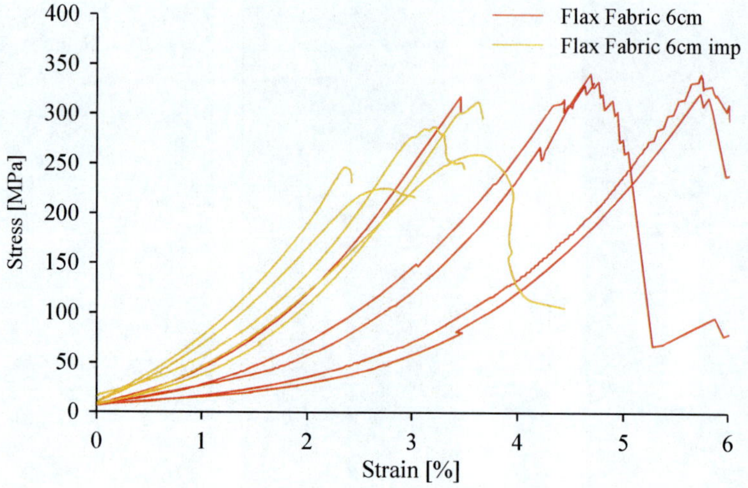

Fig. 7.18 Stress-strain response of non-impregnated and impregnated flax strips

Table 7.3 Main parameters of flax strips tested in tension

Series	Cross section	σ_{max}	ε_{max}	E
	(mm^2)	(MPa)	(%)	(GPa)
Flax Fabric 6 cm	6.12	331	4.9	12.50
Co.V. (%)	–	3	19	6
Flax Fabric 6 cm imp	6.12	266	3.1	12.43
Co.V. (%)	–	13	18	14

step characterised by a lower stiffness that increases with the strain up to achieve a quasi-linear branch before the failure. The latter represents the stiffness E reported in the table.

The series Flax Fabric 6 cm was characterised by a failure mode in which the progressive rupture of the flax threads occurred. The Flax fabric 6 cm-imp specimens failed at lower values of stress and with a mixed failure characterised by filament rupture and slipping of filaments within the coating polymer. Moreover, unlike non-impregnated fibres, the impregnated textile is also characterised by a transversal stiffness. By increasing the strain, due to the Poisson effect, the textile tends to get thinner in the middle contributing to a premature failure. Although being characterised by the same main value of stiffness in the linear branch, impregnated strips were characterised by lower values of the strain at failure. Such behaviour, in line with the evidence showed in previous paragraph where impregnated threads were tested in tension, confirmed that the coating reduces the deformability of the textile.

The tensile response of Flax TRMs in terms of load-displacement is shown in Figs. 7.19, 7.20 and 7.21. The response is also shown in terms of stress-strain in Figs. 7.22, 7.23 and 7.24.

The stresses were assessed, and reported within the graphs, with respect to both the mortar and textile cross section in order to represent both configurations in which,

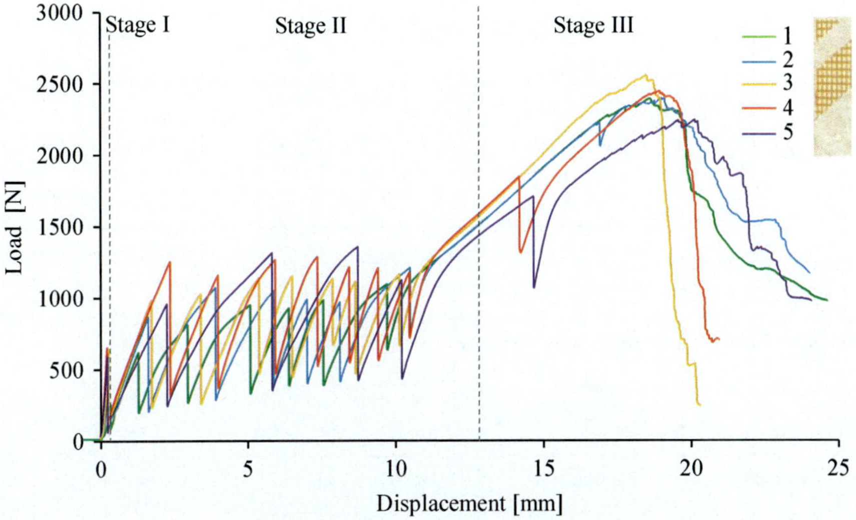

Fig. 7.19 Load-displacement curves concerning Flax TRM-2L specimens

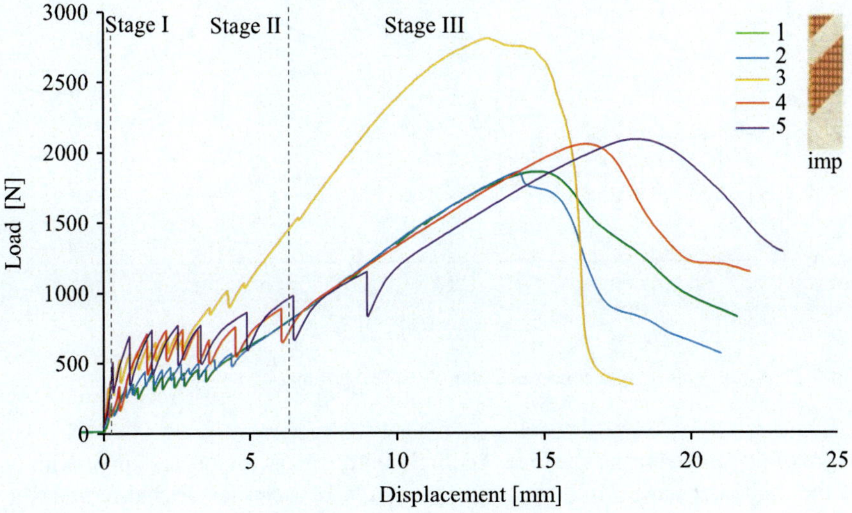

Fig. 7.20 Load-displacement curves concerning Flax TRM-2L-imp specimens

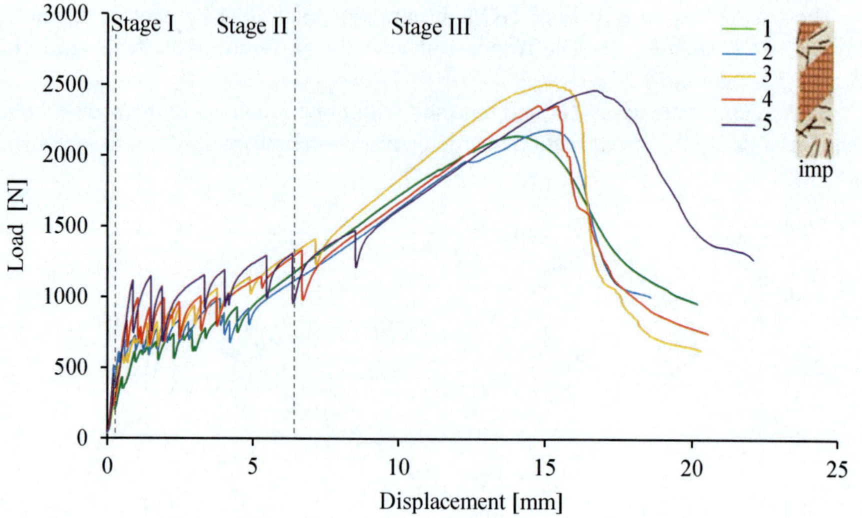

Fig. 7.21 Load-displacement curves concerning Flax TRM-2L-imp-SF specimens

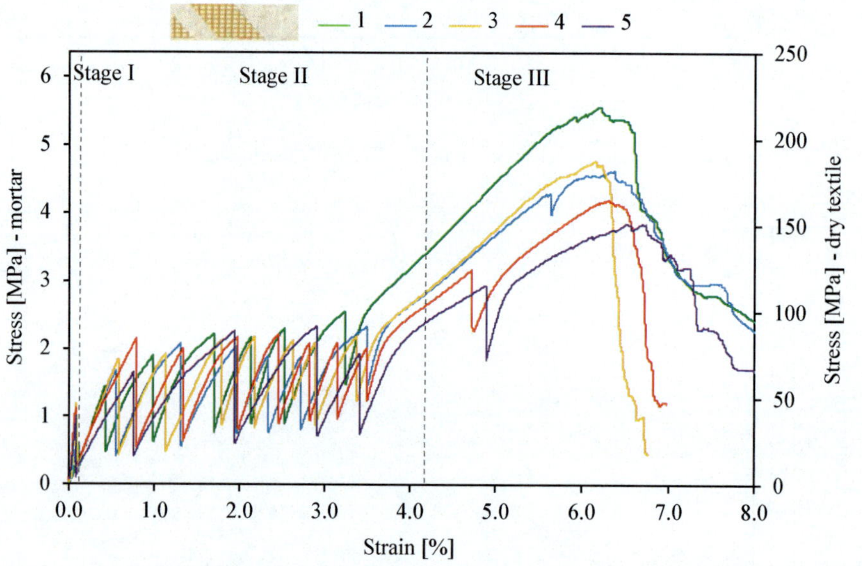

Fig. 7.22 Stress-strain curves concerning Flax TRM-2L specimens

in proximity of the cracks, the load is entirely borne by the textile, and, in proximity of the sections characterised by a full interaction between textile and mortar, the stress may be assumed uniformly distributed along the TRM cross section.

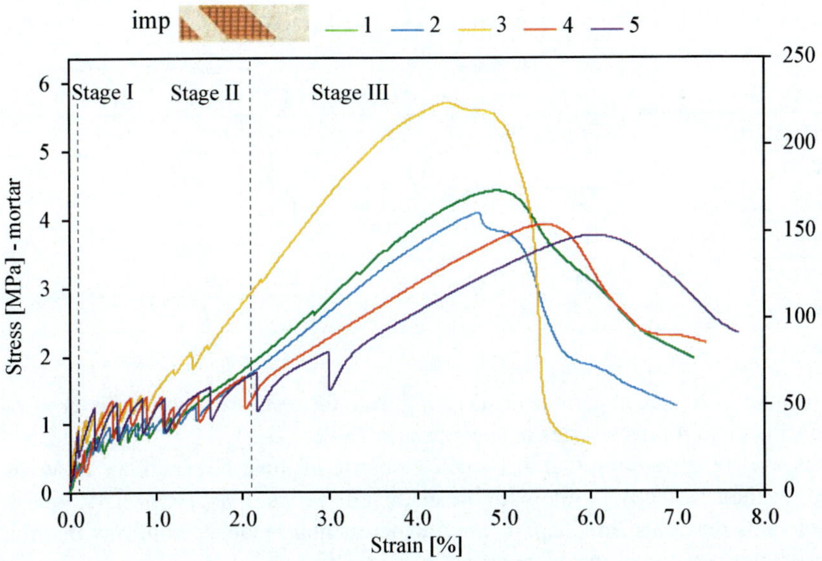

Fig. 7.23 Stress-strain curves concerning Flax TRM-2L-imp specimens

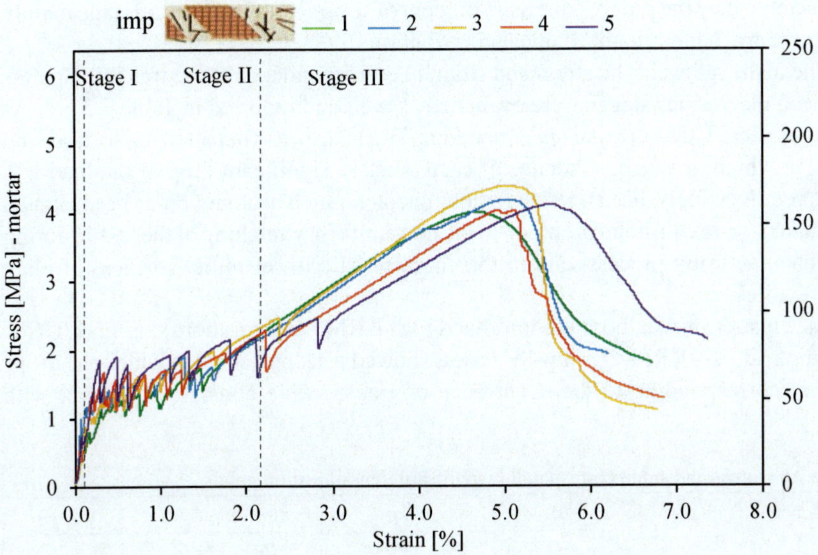

Fig. 7.24 Stress-strain curves concerning Flax TRM-2L-imp-SF specimens

Table 7.4 Main properties of Flax TRM specimens tested in tension

Series	P_{max}	$disp_{max}$	σ_{max}	ε_{max}	$\sigma_{textile}$	$\sigma_{max}/f_{textile}$
	(N)	(mm)	(MPa)	(%)	(MPa)	(%)
Flax TRM-2L	2402	19.0	5.00	6.3	196	59
Co.V. (%)	5	3	5	3	5	5
Flax TRM-2L-imp	2129	15.3	4.44	5.1	174	65
Co.V. (%)	18	13	18	13	18	18
Flax TRM-2L-imp-SF	2331	15.1	4.86	5.0	191	72
Co.V. (%)	7	6	7	6	7	7

The main mechanical parameters deriving from the tests are reported, in terms of mean value, and for each series of specimen, in Table 7.4.

It is worth highlighting that the tensile response of the reference Flax TRM-2L series resulted perfectly in line with the analogous series, Flax TRM-2L-T, whose behaviour is described in Chap. 4. Such experimental evidence confirms that the desired boundary conditions were properly reproduced.

As with as the reference series, also the others were characterised by the classical trilinear behaviour characterised by: the first elastic stage, the transition behaviour characterised by the progressive development of several cracks, the third stage mainly governed by the textile mechanical contribution.

The main values of the stress and strain in correspondence to the transition points between the various stages were evaluated as well and reported in Table 7.5.

As expected, the series of specimen Flax TRM-2L, was characterised by a second stage in which, in correspondence of each crack, a significant drop of the load was observed. As widely discussed in previous chapters, such behaviour may be attributed to a matrix-to-textile bond problem, to the non-uniform stretching of the textile during the manufacturing process, and to the characteristic lower initial stiffness of plant fibre textiles.

As a matter of fact, both the improved Flax TRM configurations, i.e. Flax TRM-2L-imp and Flax TRM-2L-imp-SF series, showed a second stage in which the drops of load corresponding to the occurrence of cracks, were considerably lower with

Table 7.5 Stress and strain corresponding to the different transition points

Specimen	σ_1	ε_1	σ_2	ε_2
	(MPa)	(%)	(MPa)	(%)
Flax TRM-2L	1.15	0.1	2.53	3.7
Co.V. (%)	24	7	6	5
Flax TRM-2L-imp	0.68	0.1	1.75	2.0
Co.V. (%)	48	25	37	40
Flax TRM-2L-imp-SF	0.96	0.1	2.34	2.1
Co.V. (%)	11	40	17	26

respect to what observed in the reference series. Moreover, it can be observed that the number of cracks exhibited in the two improved configurations, was higher with respect to the reference series. Such aspect, emphasized that the fibre treatment caused alterations in the fibre-to-mortar bond properties, and in the textile stiffness, that resulted in a higher capacity to transfer the stress from the textile to the matrix, with a consequent reduction of the distance between consecutive cracks.

With respect to the third stage, although dry and impregnated textile tested in tension exhibited the same stiffness, the average value of the slope of the curves concerning the improved Flax TRMs (series Flax TRM-2L-imp and Flax TRM-2L-imp-SF) was of about 10% lower than what observed in the reference series. Such experimental evidence may be due to the higher number of cracks occurred in the improved Flax TRMs.

The specimen Flax TRM-2L-imp-3 was characterised by a much stiffer, and with a higher strength, response, with respect to the other specimens. Such behaviour can be attributed to the impregnation treatment adopted. Specifically, the impregnation treatment was manually performed, and applied on several textile portions. Such procedure, adopted without having at disposal standardized industrial techniques, unavoidably resulted in a heterogeneity among the textile strips. In order to reduce the influence of such variability in the composite behaviour, the TRM coupons were performed by uniformly distributing the flax strips deriving from the different portions, among the various specimens and series. The better performance exhibited by the specimen Flax TRM-2L-imp-3 may be attributed to the use of flax strips characterised by an impregnation treatment occurred in a proper manner. For the same reason the response of the series in which impregnated textiles were adopted, is characterised by a significant variability (see coefficient of variation of the mechanical parameters reported in Table 7.4).

By comparing the stress at the transition point between the Stages I and II, σ_1, it can be noticed that the average value of the reference series was higher with respect to the others. Such evidence may be attributed to the fact that the less significant contribution of the textile in the elastic phase let the specimen to be closer to the mortar direct tensile conditions. However, by comparing the mean values of the stress σ_1 of the improved Flax TRMs, it is observed that the series Flax TRM-2L-imp-SF was characterised by a value of about the 30% higher than those of Flax TRM-2L-imp series. This aspect is clearly shown in Fig. 7.25, where representative specimens of each series were compared.

The comparison shows that the load values corresponding to each crack opening, is generally higher in improved Flax TRMs with respect to the reference series. Moreover, by comparing the Flax TRM-2L-imp and the Flax TRM-2L-imp-SF it is observed that the latter was characterised by an average value of the stress within the second stage, much higher than the one observed in the other series. This behaviour is due to the presence of the curaua short fibres that increased the tensile strength of the mortar. This aspect is of fundamental importance in view of applications of the TRM as reinforcement of structural elements. In fact, the higher stress values on the second stage, and the lower losses of strength corresponding to the crack openings, may lead toward a hardening post-elastic behaviour.

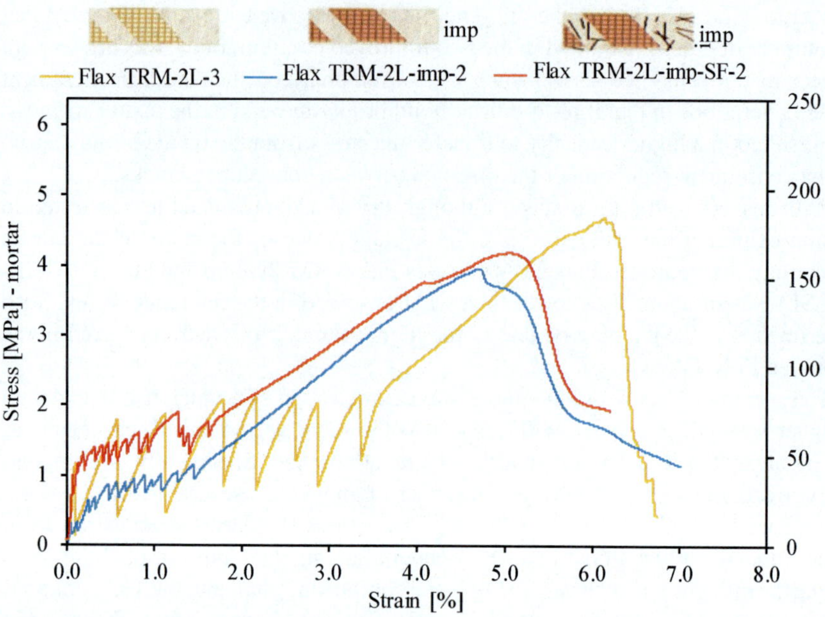

Fig. 7.25 Comparison between representative specimens of the series tested in tension

In terms of ε_2, the use of impregnated textiles led to an evident reduction of the deformability within the second stage, with a mean value of the strain that decreased of about the 50% in the improved TRMs with respect to the reference series. The latter represents a fundamental aspect. In fact, as known, once applied on a structural element, the TRM composite is considered to confer an actual contribution when behaving in the third stage. Therefore, the textile impregnation treatment leads to a TRM system capable to exhibit its contribution, once applied on structural element, at smaller values of deformation, with respect to the reference Flax TRM. The smaller deformability of the composite may correspond, on the structural element scale, to a post elastic behaviour characterised by less significant abrupt losses of strength.

The failure mode was characterised, in all specimens, by a progressive development of cracking along the free length, up to the rupture of the textile in proximity of a crack (Fig. 7.26d).

The comparison between the cracked representative specimens of the three series, shows that Flax TRM-2L-imp and Flax TRM-2L-imp-SF specimens were characterised by a higher number of cracks, closely spaced with respect to the reference specimen (Fig. 7.26a, b and c).

With respect to the maximum strength achieved, as expected considering the tensile strength exhibited by non-impregnated and impregnated textile strips, the reference series was characterised by higher values of the maximum stress. However, in terms of exploitation ratio, σ_{max}/f_{tex} (representing the ratio between the maximum stress achieved in the composite and the maximum strength achievable in the textile),

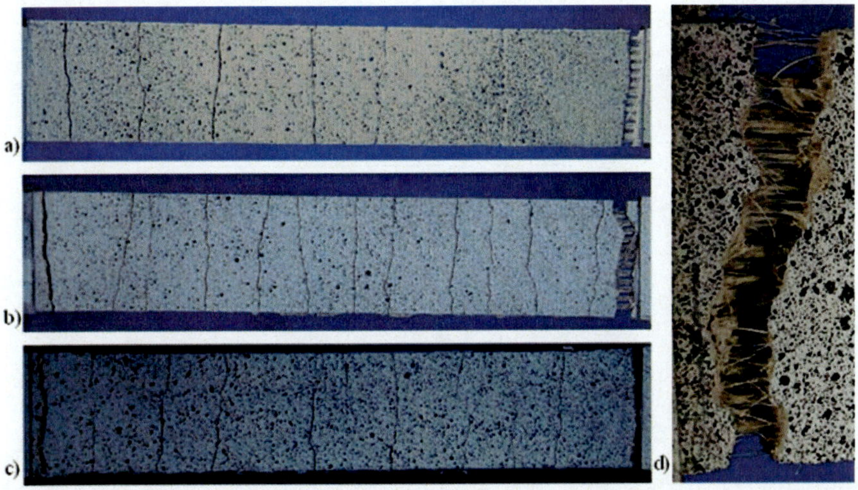

Fig. 7.26 Failure mode of representative specimens: **a** Flax TRM-2L-3; **b** Flax TRM-2L-imp-2; **c** Flax TRM-imp-SF-2; **d** Rupture of the textile

a significant increase of 10 and 20% was observed with respect to Flax TRM-2L-imp and Flax TRM-2L-imp-SF series.

7.2.3 Analysis of the Crack Pattern

The analysis of the crack pattern was carried out on a representative specimen of each series tested in tension, i.e. Flax TRM-2L-3 (also referred to as reference Flax TRM), Flax TRM-2L-imp-2 and Flax TRM-2L-imp-SF-2 specimens (also referred to as improved Flax TRMs).

In line with what was proposed in paragraph 4.1.4, DIC images were treated in order to construct the displacement field of the external surface of the specimens, over a length of 200 mm in the middle of the free length of the specimens. For each specimen significant images were selected, corresponding to the occurrence of cracks, the transition points between different stages, and the cracked specimen at maximum.

The displacement over a longitudinal section corresponding to the selected images, for each representative specimen, is shown in Figs. 7.27, 7.28 and 7.29.

The image on the top of the figures represents the displacement field of the external surface of the TRMs, over a length of 200 mm, in correspondence to the achievement of the maximum strength.

By comparing the graphs, it is clear that the improved Flax TRMs were characterised by a much higher number of cracks developed during the test. Moreover, such cracks, developed at much lower values of the strain. Such behaviour confirms, from

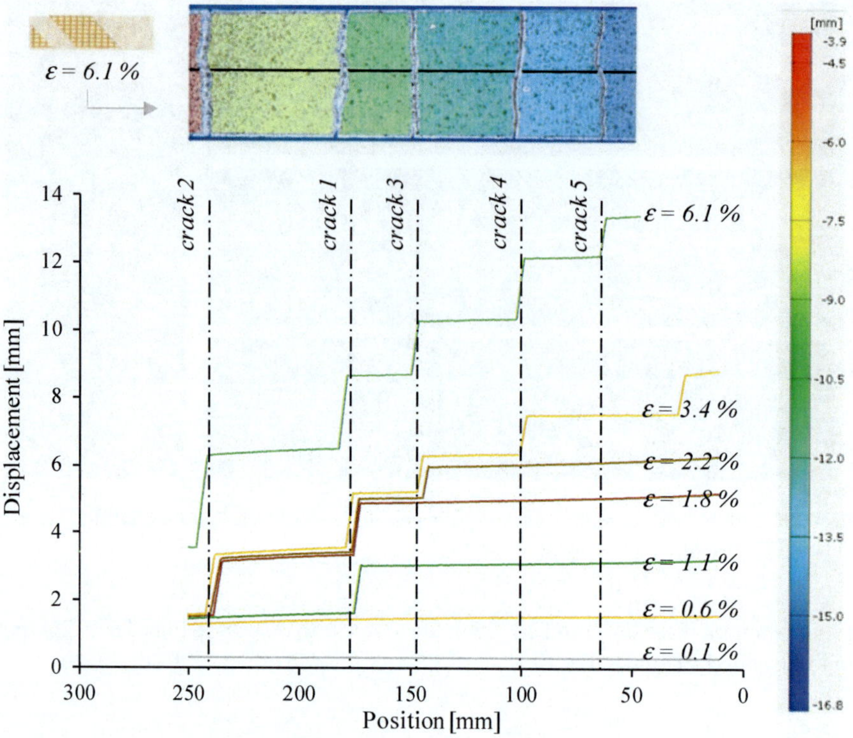

Fig. 7.27 Displacement field of the middle longitudinal section (Flax TRM-2L-3)

another perspective, the deformability of the improved Flax TRMs in the second stage, with respect to the reference TRM.

The capacity of developing cracks at lower values of strain, is clearly highlighted in Fig. 7.30, where the evolution of the mean crack spacing as a function of the axial strain, is shown for the three representative specimens. The curves are a logarithmic approximation of discreet values. This representation clearly emphasises the best mechanical performance of the improved Flax TRMs with respect to the reference coupon.

By analysing the displacement fields it is also possible to see that the improved Flax TRMs were characterised by shorter distances between consecutive cracks, as well as smaller crack openings.

The mean values of the distance between consecutive cracks, $\delta_{0,mean}$, and the maximum crack opening, w_{max}, concerning the representative specimens are listed in Table 7.6. The maximum value of the crack opening resulted up to 3 times smaller in improved Flax TRMs with respect to the reference behaviour. Such aspect is fundamental in view of application of the composite as external reinforcement of structural elements. In fact, the closer space between cracks reduces the contact of

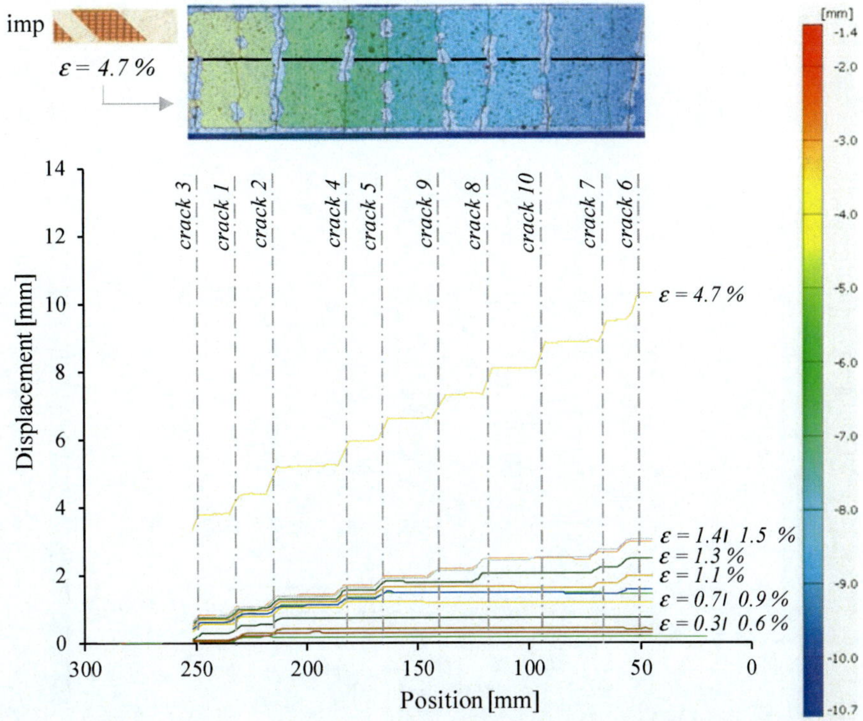

Fig. 7.28 Displacement field of the middle longitudinal section (Flax TRM-2L-imp-2)

the textile with the external environment, improving the capacity to resist to ageing processes.

With respect to the approximated analytical model proposed in paragraph 4.1.4, starting from the values of the mean spacing between consecutive cracks, the tensile strength of the mortar and the perimeter of non-impregnated and impregnated textile cross section, an approximated estimation of the maximum interfacial shear stress between the fabric and the mortar was carried out. The mean value for each representative specimen is reported in Table 7.6. Although being an approximated estimation, this parameter provides interesting information.

The mean shear stress value concerning the reference series is in line with what observed in Chap. 4, confirming once again the desired conditions were well reproduced. In both of the improved TRM configurations a significant increase of the mean shear stress was observed with respect to the reference series. This result is not in line with what observed in the previous paragraph where the bond between the mortar and non-impregnated and impregnated threads was analysed by means of pull-out tests. Such experimental investigation showed that the impregnation caused a reduction in the matrix to fibre bond capacity.

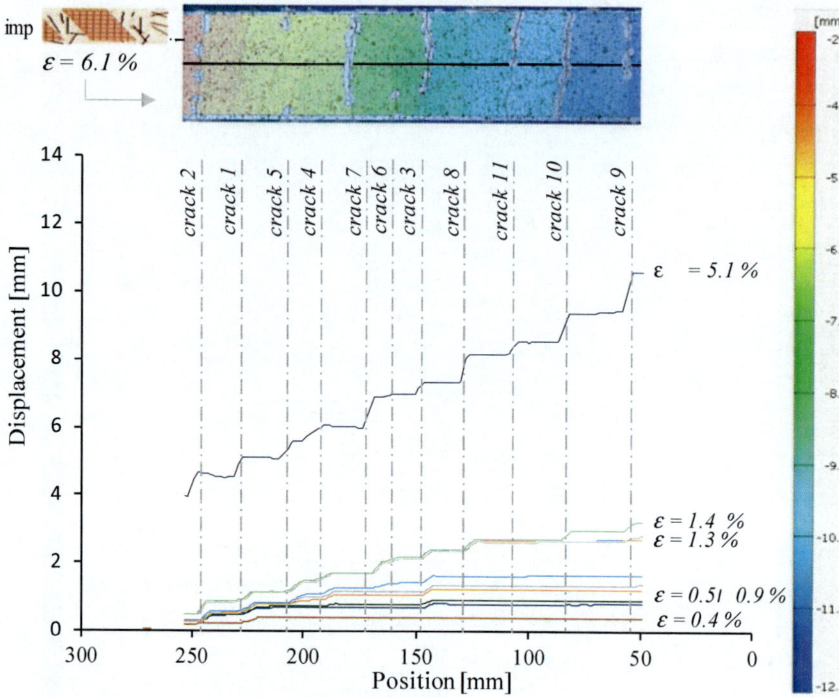

Fig. 7.29 Displacement field of middle longitudinal section (Flax TRM-2L-imp-SF-2)

This discrepancy may be due to two aspects. The approximated model, by considering constant the shear stress along the length, did not take into account the stiffness of the textile, that in fact plays an important role in the mechanical behaviour of the system. The other aspect is related to the fibre impregnation treatment. As a matter of fact, the impregnation of the single threads, did not reproduce the same conditions created by impregnating portions of textile. Specifically, the removal of the coating substance in excess from the flax thread was much easier than removing the polymer form the bidirectional fabric. As a result, the textile strips were characterised by a higher amount of coating substance with respect to the flax threads.

This aspect emphasizes that, although showing very satisfactory results in terms of mechanical behaviour of the composite, the impregnation procedure adopted still needs to be refined, perhaps by adopting industrial techniques, and further studies are needed to define the best coating substance. Moreover, for further investigation, it would be more reliable to extract flax threads to submit to pull-out tests, directly from the impregnated textile. By doing so it would be easier to reproduce, by means of pull-out tests, the bond conditions of composite tensile tests.

Although having a similar mean value of the distance between cracks, the Flax TRM-2L-imp-SF representative specimen showed a mean shear stress of about the 30% higher than the one exhibited by the Flax TRM-2L-imp representative specimen.

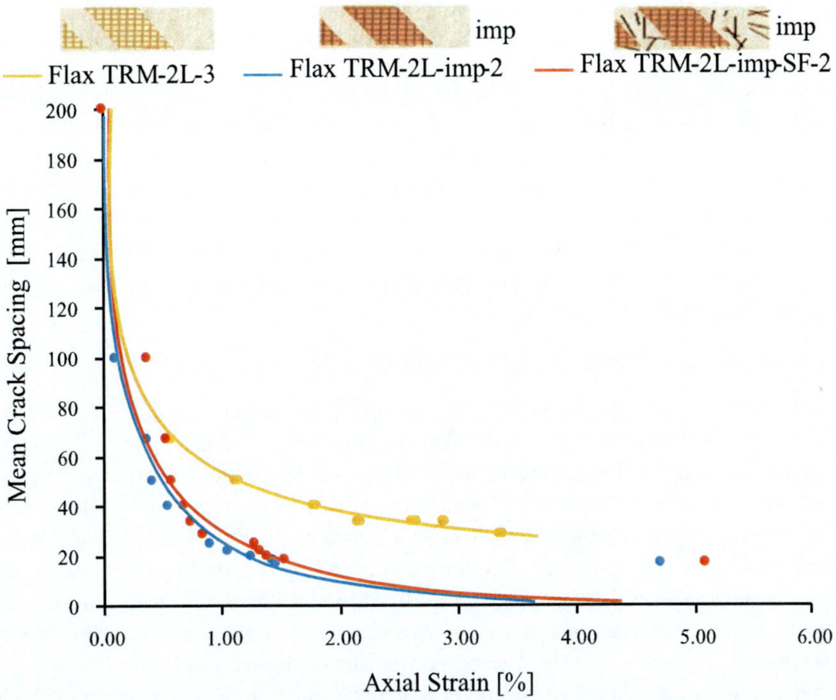

Fig. 7.30 Mean crack spacing as a function of the strain for representative specimens

Table 7.6 Main parameters of the analysis of cracks of representative specimens

Series	Spec.	$\delta_{0,mean}$	w_{max}	τ
		(mm)	(mm)	(MPa)
Flax TRM-2L	3	43	2.70	0.22
Co.V. (%)		29	–	25
Flax TRM-2L-imp	2	23	0.83	0.57
Co.V. (%)		31	–	28
Flax TRM-2L-imp-SF	2	27	1.12	0.74
Co.V. (%)		28	–	28

Such result is due to the higher tensile strength of the short-fibre reinforced mortar, that, according to the model, is directly proportional to the shear stress. However, although the approximated nature of the analytical model, it is reasonable to assert that the addition of short fibres within the mortar may result in an increased capacity of transferring the stress from the textile to the matrix, promoting an homogeneous distribution of the stress over the entire composite cross section.

7.3 Conclusions

The aim of the chapter is to show that by applying some of the techniques available in literature, it is possible to improve the mechanical performance of plant based TRM composite systems. At first, the influence of a fibre impregnation treatment on the morphology and tensile strength of flax threads, and on the bond behaviour with a hydraulic lime-based mortar, was investigated. Secondly, the tensile response of improved TRMs, obtained by means of the application of impregnated textiles, and by means of the addition of curaua short fibres within the mortar, was investigated and compared with reference TRMs.

The main findings of the experimental investigation are listed as follows:

- as shown by analysing SEM images, the coating treatment significantly changed the morphology of the flax threads. The FIERr decreased of about the 35% as well as the variability of the geometric properties, with a much lower dispersion in the measurements with respects to what observed with non-impregnated fibres;
- the impregnation treatment resulted in a favourable effect in terms of tensile behaviour of the flax threads. Despite the tensile strength slightly decreased with the impregnation, the deformability significantly decreased, with an ultimate strain much smaller than the one observed by testing non-impregnated threads, due to a stiffer response in the initial branch of the stress-strain curve;
- the coating treatment led to the creation of a polymeric weak layer between fibres and matrix resulting in a reduction of the maximum fibre-to-mortar interface shear stress of about the 30%;
- the impregnation of flax textile strips had a favourable effect on the tensile behaviour. In fact, although characterised by slightly lower strength values, impregnated textiles exhibited a lower deformability by achieving the linear stiffness at lower strains with respect to non-impregnated textiles;
- both the improved TRMs proposed in the study showed a second stage in which the typical drops of load corresponding to the occurrence of a crack in plant-based composites, were considerably lower with respect to what observed in the reference TRM;
- the use of impregnated flax textiles as reinforcement of TRMs significantly reduced the deformability of the system in the second stage, with a main value of the transition strain ε_2 reduced of about the 50% with respect to the reference values;
- the addition of short fibres within the mortar resulted in a mean value of the stress within the second stage, significantly higher than the value observed in the improved TRMs conceived without improving the mortar;
- the improved TRMs failed assuming higher values of the exploitation ratio with respect to the reference series, behaving as a more efficient composite material;
- the analysis of the crack pattern showed that the improved TRMs were characterised by a much higher number of cracks developed at much lower values of strain with respect to the reference behaviour;

– the application of the improvement techniques resulted in lower cracks opening, reducing the contact with the external environment, hence letting hope to an improved durability performance.

It is worth emphasising that the showed results represent a preliminary study, and the impregnation procedure needs to be standardised in order to make the results more reliable (at present they are characterised by a considerable coefficient of variation). The experimental evidence showed within the chapter confirmed that the application of simple improving techniques can significantly improve the mechanical performance of plant based TRMs. Moreover, according to the critical observation reported in Chap. 6, these improvements may correspond to a much better performance of the composite system once applied to structural elements to enhance their strength. The lower deformability of the TRM in the cracked zone, the less significant loss of strength corresponding to the crack development, a more efficient behaviour as composite characterised by a wider development of cracks through the entire specimen, and the higher values of the stresses in the cracked zone, may possibly result in a more satisfying mechanical behaviour of externally strengthened masonry elements, perhaps by achieving higher elastic capacity, and an hardening post-elastic behaviour.

Finally, it is worth highlighting that the improvement solutions adopted, were not selected through sophisticated technical analysis, but were rather chosen with respect to the ease availability of the materials employed, and to few encouraging similar literature evidences. Nevertheless, the scientific outcomes resulted more than promising. Reasonably, further investigations aimed at optimizing improvement techniques, may only result in a mechanical behaviour even more satisfactory.

References

1. Cevallos OA, Olivito RS (2015) Effects of fabric parameters on the tensile behaviour of sustainable cementitious composites. Compos Part B Eng 69:256–266
2. EN 1015-3:1999 (1999) Methods of test mortar for masonry. determination of consistence of fresh mortar (by Flow Table); European committee for standardization. Brussels, Belgium
3. EN 196-1:1994 (1994) Methods of testing cement—part 1: determination of strength. European committee for standardization. Brussels, Belgium
4. Ferrara G, Martinelli E (2018) Tensile behaviour of textile reinforced mortar composite systems with flax fibres. In: Proceedings of the 12th fib International Ph.D. Symposium in Civil Engineering, pp 863–869
5. Ferrara G, Caggegi C, Martinelli E, Gabor A (2020) Shear capacity of masonry walls externally strengthened using Flax-TRM composite systems: experimental tests and comparative assessment. Constr Build Mater 261:120490
6. Ferrara G, Pepe M, Martinelli E, Toledo Filho RD (2019) Influence of an impregnation treatment on the morphology and mechanical behaviour of flax yarns embedded in hydraulic lime mortar. Fibers 7(4):30
7. Ferreira SR, de A, Silva F, Lima PRL, Toledo Filho RD (2015) Effect of fiber treatments on the sisal fiber properties and fiber–matrix bond in cement based systems. Constr Build Mater 101:730–740

8. Ferreira SR, Martinelli E, Pepe M, de Andrade Silva F, Toledo Filho RD (2016) Inverse identification of the bond behavior for jute fibers in cementitious matrix. Compos Part B Eng 95:440–452
9. Jia Y, Fiedler B (2018) Influence of furfuryl alcohol fibre pre-treatment on the moisture absorption and mechanical properties of flax fibre composites. Fibers 6(3):59
10. Zukowski B, de Andreade Silva F, Toledo Filho RD (2018) Design of strain hardening cement-based composites with alkali treated natural curaua fiber. Cement Concrete Compos 89:150–159
11. Zukowski B, dos Santos ERF, dos Santos Mendoça YG, de Andreade Silva F, Toledo Filho RD (2018) The durability of SHCC with alkali treated curaua fiber exposed to natural weathering. Cement Concrete Compos 94:116–125

Chapter 8
Conclusions

The Thesis was intended at assessing the potential of Flax TRM composite systems as external strengthening system of masonry structural elements. It order to meet the need of a deep understanding of the mechanical behaviour of plant fibre based TRMs, the study proposes a comprehensive characterisation that moves from the analysis of the microstructure of the flax threads constituting the textile, up to the investigation of the efficiency of the composite system as reinforcement of masonry walls. The main outcomes of the research study are discussed as follows.

8.1 Main Outcomes

As preliminary investigation, the physical and mechanical properties, as well as some durability aspects, of the flax textile adopted in the study were analysed. The results demonstrated that the flax fabric under investigation, with promising mechanical properties can be utilised as internal reinforcement for TRM systems. In terms of durability the fibres showed a good performance in the short term, and significant decreases of strength in accelerated ageing exposures, highlighting the need of specifically designed treatments to cope with this issue.

In light of the encouraging outcomes on the textile scale of analysis, the study moved to the investigation of Flax TRM composite systems. Tensile tests showed that the response in tension was characterised by a trilinear behaviour, differing from the typical behaviour exhibited by conventional high strength fibre based TRMs, in the exhibition of significant drops of the stress in correspondence to crack occurrence. Such experimental evidence may be due to the low initial stiffness of plant fibres when subjected to tensile loading conditions. However, the specimens exhibited a promising behaviour with a fibre exploitation ratio of about the 70%. With respect to the Flax TRM-to-masonry substrate shear bond behaviour, the experimental evidences showed a rupture of the textile in the free length of the charging edge. This behaviour emphasised that the system is characterised by a good adhesion

G. Ferrara, *Flax-TRM Composite Systems for Strengthening of Masonry*, Springer Theses,
https://doi.org/10.1007/978-3-030-70273-1_8

at the fibre-to-mortar interface with respect to the tensile strength of the flax textile itself. The experimental investigation carried out to analyse the durability of Flax TRM composites showed that the system, exposed to a conventional ageing protocol, complies with the acceptance criteria proposed by standards in the field.

Based on the satisfactory experimental evidences deriving from a characterisation at the composite scale of analysis, the reinforcement systems were applied to masonry structural elements to assess their efficiency in enhancing the in-plane capacity of the walls. The application of the Flax TRM system significantly affected the mechanical response of the masonry elements tested in diagonal compression, by conferring them an increase in shear capacity, and a ductile behaviour completely absent in unreinforced elements. The experimental evidences also showed a significant loss of strength corresponding to the cracking of the wall in proximity of the maximum capacity, due to the significant deformability of the reinforcing flax textile at low values of strength. A comparative analysis with works available in the literature showed that the study is in line with similar investigations and is characterised by a low fibre reinforcement ratio. It was also showed that increasing the fibre amount by overlapping different textile layers not necessary ends up in a more performant behaviour.

A critical analysis of the strain fields of the TRM system, both in tensile and diagonal compression tests, showed that there is a relationship between the behaviour of the composite in the two different configurations. By analysing such correspondence, it was possible to define in which terms a more satisfactory tensile behaviour may lead to a better performance on the structural scale of analysis. Specifically, with respect to TRM tensile response, an increase of the mean stress in the second stage, a reduction of the loss of strength corresponding to the crack openings, an increase of the number of cracks developed at smaller strain levels, would probably results in a composite system that, once applied as reinforcement of structural elements, may confer them a behaviour tending toward a higher elastic capacity, and an hardening post-elastic behaviour.

Finally, with the aim of going further than merely confirm the promising potential of the proposed reinforcement system, the study proposed the application of technical solutions with the purpose of showing that there is still room for improvements in the mechanical behaviour of plant based TRM composites. Specifically, by means of Flax TRM tensile tests, the efficiency of a textile impregnation treatment, and of the addition of curaua short fibres within the mortar, was investigated. The use of impregnated flax textile resulted in a satisfactory improvement of the mechanical response of the system. In fact, although characterised by slightly lower strength, the improved TRMs exhibited a second stage with drops of the load corresponding to the occurrence of cracks, significantly lower with respect of the reference behaviour. Moreover, the application of the impregnated textiles considerably reduced the deformability of the system with a main value of the transition strain, ε_2, reduced of about the 50% with respect to the reference behaviour. Furthermore, the addition of short fibres within the mortar resulted in a main value of the stress within the second stage, significantly higher than the value observed in the improved TRM conceived without improving the mortar. As a matter of fact. the experimental evidences showed that the application of simple available techniques can significantly improve the mechanical performance of plant based TRMs.

8.2 Perspectives of the Study

The study, presenting a large amount of results, contributes to advancing knowledge about the use of plant textiles as reinforcement in mortar-based composites, and paves the way for further research activities aimed at optimising the sustainable strengthening system object of study. Many perspectives of the study may be defined:

- further mechanical investigations aimed at the definition of optimised improvement techniques, such as: performant textile treatments by means of chemical and/or physical procedures, the use of efficient coating substances improving the textile response and the fibre-to-mortar bond, the use of optimised fibre reinforcement amounts and textile arrangements, efficient matrix modifications improving the composite performance;
- comprehensive analysis on the durability of plant based TRMs, with respect to significant ageing environments, and by defining fibre treatments able to improve such aspect;
- multicriterial studies aimed at providing an exhaustive assessment of the sustainability of these types of reinforcement systems, and at comparing it with conventional composite systems nowadays largely adopted;
- the definition of actual possible applications of the studied composite system, according to its mechanical performance, as a reinforcement technique for structural elements and for elements of secondary importance.

Future research studies moving in the according to these directives, may end up in optimisation of the strengthening technique by means of plant based TRM composites, hence hopefully in a significant diffusion of these types of sustainable reinforcement systems in actual civil engineering applications.

Appendix

Displacement field of the middle longitudinal section of specimens of the series Flax TRM-1L-T and Flax TRM-2L-T.

See Figs. A.1, A.2, A.3, A.4, A.5, A.6, A.7, A.8 and A.9 .

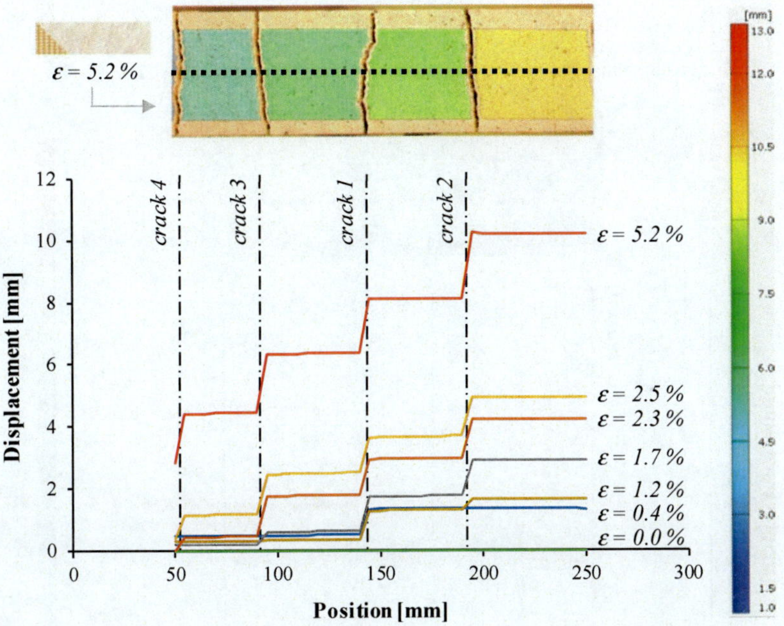

Fig. A.1 Displacement field of middle longitudinal section (specimen Flax TRM-1L-1)

G. Ferrara, *Flax-TRM Composite Systems for Strengthening of Masonry*, Springer Theses, https://doi.org/10.1007/978-3-030-70273-1

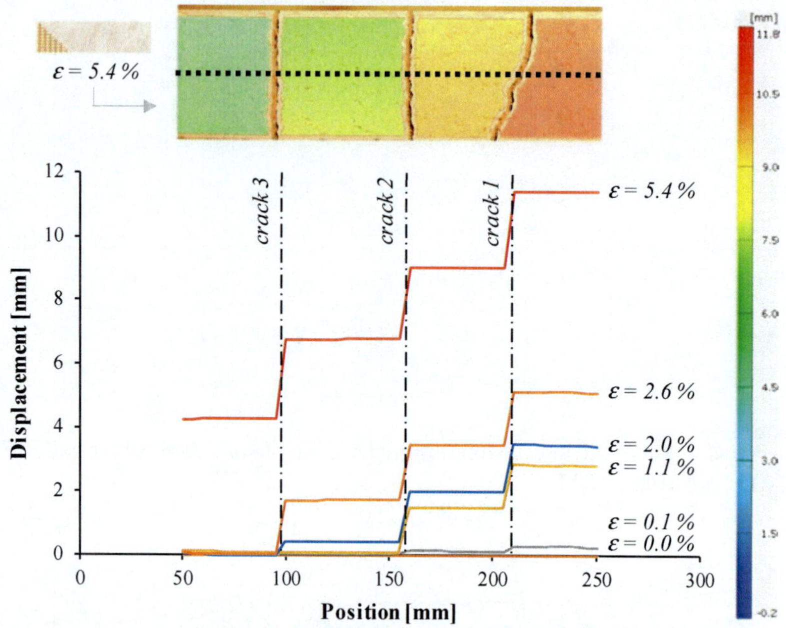

Fig. A.2 Displacement field of middle longitudinal section (specimen Flax TRM-1L-2)

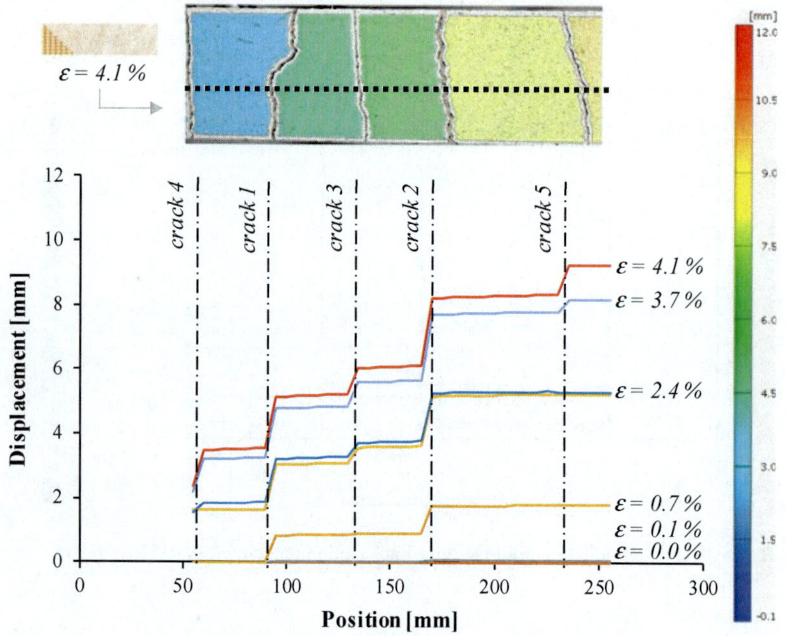

Fig. A.3 Displacement field of middle longitudinal section (specimen Flax TRM-1L-3)

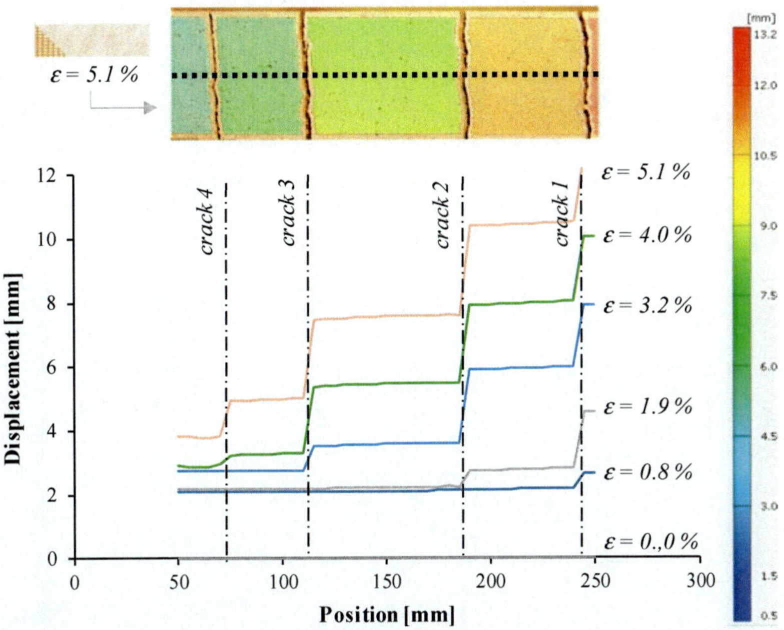

Fig. A.4 Displacement field of middle longitudinal section (specimen Flax TRM-1L-4)

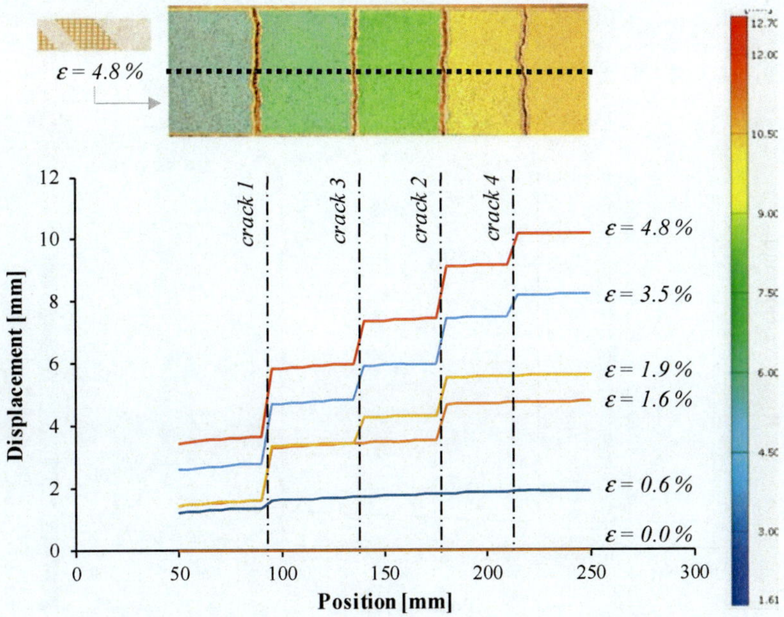

Fig. A.5 Displacement field of middle longitudinal section (specimen Flax TRM-1L-4)

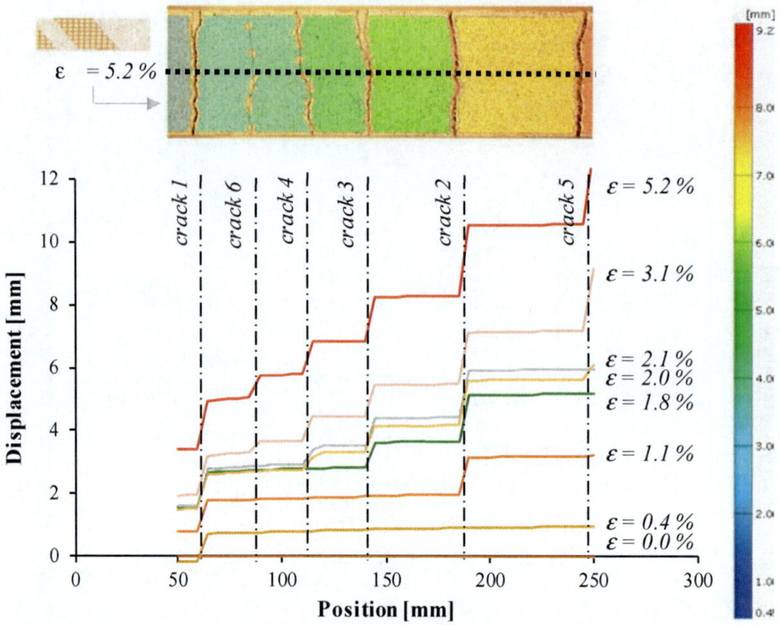

Fig. A.6 Displacement field of middle longitudinal section (specimen Flax TRM-2L-2)

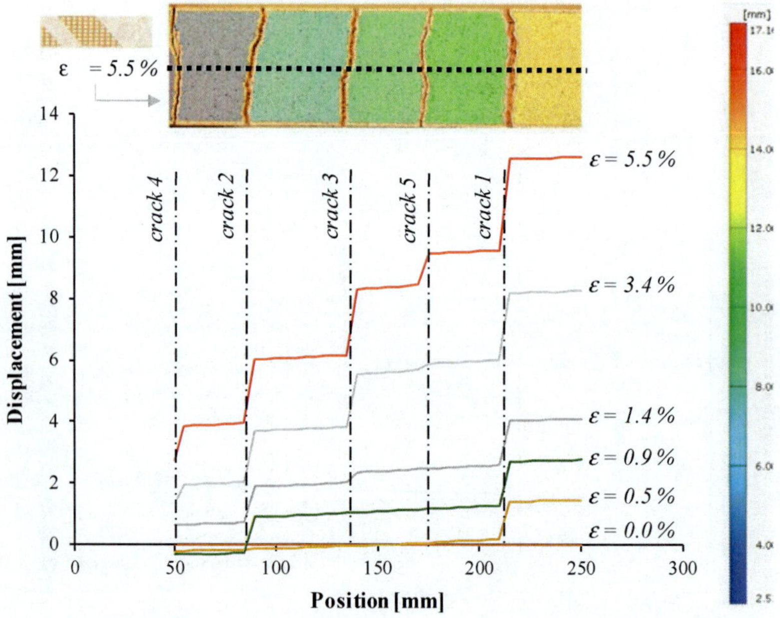

Fig. A.7 Displacement field of middle longitudinal section (specimen Flax TRM-2L-3)

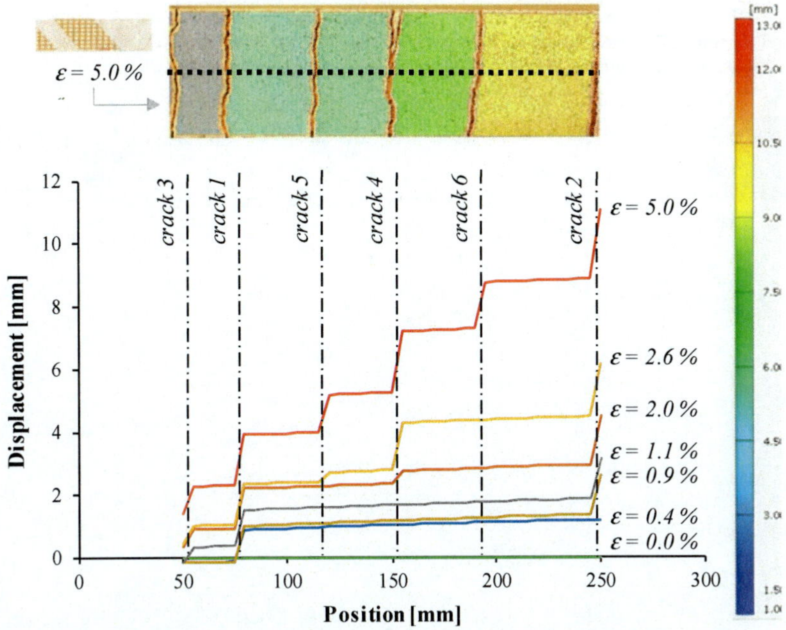

Fig. A.8 Displacement field of middle longitudinal section (specimen Flax TRM-2L-4)

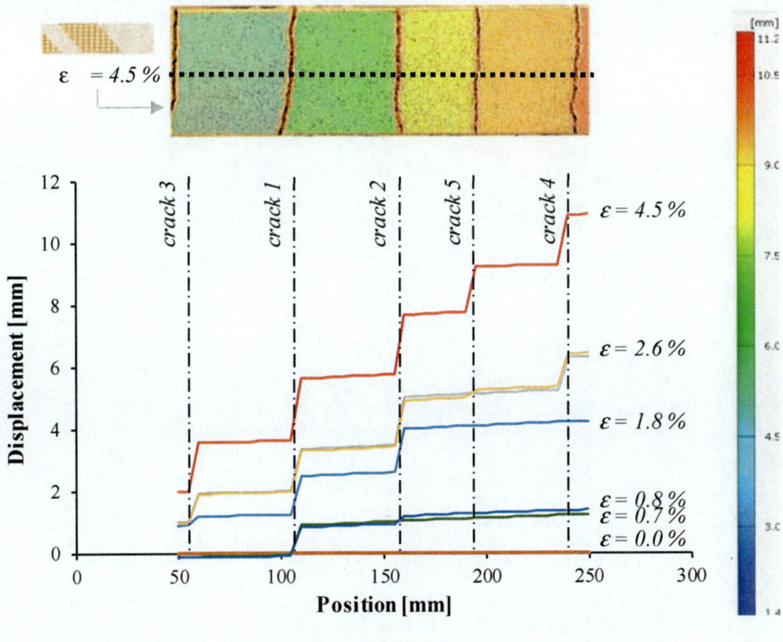

Fig. A.9 Displacement field of middle longitudinal section (specimen Flax TRM-2L-5)

La borsa di dottorato è stata cofinanziata con risorse del
Programma Operativo Nazionale Ricerca e Innovazione 2014-2020 (CCI 2014IT16M2OP005),
Fondo Sociale Europeo, Azione I.1 "Dottorati Innovativi con caratterizzazione Industriale"

UNIONE EUROPEA
Fondo Sociale Europeo

Ministero dell'Istruzione
dell'Università e della Ricerca

PON
RICERCA
E INNOVAZIONE